In

The Next Dimension

The Book Of Phi. Volume 2

(The 17th Vedic Mathematics Sutra)

Jain

Bridging
- the Phi Ratio of 1:1.618033
- Vedic Mathematics
- Magic Squares
- Platonic Solids and the
- Tetrahedral Numbers.

2003

In
The
Next
Dimension

The Book of Phi   JAIN   VOLUME 2

The 17th Vedic Mathematics Sutra

Bridging: Phi, Vedic Mathematics, Magic Squares,

Platonic and Archimedean Solids and the Figurate Number Series.

PUBLISHED BY

F. R. E. E. D. O. M. S.
an acronym for:

For Research Expressing Essential Data Of Magic Squares

For Further correspondence, contact: phone: 61 - 2 - 66844409

JAIN

777 Left Bank rd Mullumbimby Creek, 2482, Far Northern N.S.W. Australia

email: **jain@jainmathemagics.com**   **www.jainmathemagics.com**

2003

ISBN
0 9 5 9 4 1 8 0 - 2 - 4

THIS COMPENDIUM FOCUSSES ON
3 MAJOR MATHEMATICAL REVELATIONS OR ORIGINAL FORMULAE BY JAIN:

1)– THE DISCOVERY OF A MAJOR PATTERN RUNNING THROUGH PHI
(1: 1.618033) BASED ON THE DIGITAL COMPRESSION
OR REDUCED SUMS OF THE FIBONACCI SERIES THAT HAS
A DISTINCT PERIODICITY OF 24 RECURRING DIGITS OR 12 PAIRS OF 9:

0, 1, 1, 2, 3, 5, 8, 4, 3, 7, 1, 8
9, 8, 8, 7, 6, 4, 1, 5, 6, 2, 8, 1

(This was revealed in detail in Chapter 7 of THE BOOK OF PHI, Volume 1, 2002).

2)– ALL KNOWLEDGE OF THE MAGIC SQUARE'S MAGIC SUMS, KNOWN AS THE MAGIC SQUARE CONSTANTS (15, 34, 65, 111, 175, 260, 369 ETC) IS ENCODED IN THE TETRAHEDRON, SPECIFICALLY IN THE TRIADS OF THE TETRAHEDRAL SPHERE-PACKING NUMBER SERIES (1, 4, 10, 20, 35, 56, 84, 120 ETC).

3)– A NEW, BETTER AND SIMPLE FORMULA FOR "e" = 2.71828182845904523536028
THE EXPONENTIAL FUNCTION THAT IS NATURE'S NUMBER OF GROWTH AND DECAY,
USING SUCCESSIVE AND "SIMILAR POWERS", IN THE NEXT DIMENSION

eg: $$\frac{9\times9\times9\times9\times9\times9\times9\times9\times9}{8\times8\times8\times8\times8\times8\times8\times8} - \frac{8\times8\times8\times8\times8\times8\times8\times8}{7\times7\times7\times7\times7\times7\times7} = e$$

(ie: 9 to the 9th Power divided by 8 to the 8th Power, minus 8^8 divided by 7^7).

CONTENTS

PAGE	PART	SUBJECT
6	PART 1A.	• Determining the Value of Φ (Phi) Using Only Vedic Mathematics.
27	PART 1B.	• Determining the Dual Roots of Φ. (via Internal and External Division of Unity's Square).
37	PART 2.	• More Notes on the Phi Code (12 Pairs of 9) Incorporating Archimedean Solids, 4th Dimensional Pentatopes and HyperCubes.
47		• Phi and the Sutras.
50		• Our Genetic Spherical Memory of the Tetrahedron and the Cube. (Section 1 of 2).
61-67		• Includes Chris Illert's Micro-Psi Platonic Solid Symmetry of the Elements.
67-69		• Magic Square of 3x3 Creating the Atomic Structure of Beryllium.
70		• More Notes on Phi.
76		• Magic Squares and HyperCubes.
78		• Our Genetic Spherical Memory of the Tetrahedron and the Cube. (Section 2 of 2).
83		• The Derivation of Phi using Graphs.
86		• The Phi Code of 24 Recurring Digits and Rodin's Coil.
98		• References to '24' in the 4th Dimension.
102	PART 3	• Two Important Original Mathematical Discoveries by Jain
103	PART 3A	• The 17th Sutra. Linking of the Magic Square Constants to the Triplets of the Sphere Packing Numbers of the Tetrahedron.
105	Section 1:	• What are the Magic Square Constants?
115	Section 2:	• The Squared Numbers. The Cubic Numbers.
120	Section 3:	• The Triangular Numbers.

CONTENTS

PAGE	PART	SUBJECT
123	Section 4:	• The Tetrahedral Numbers.
132		– The Magic Tetrahedron
133		– Pascal's / Halayudha's Triangle (a.k.a 'The Chinese Triangle')
135		– Tetrahedral Micro-Fossils of Spores and Pollen.
136		– The Oracle of Delphi.
137	Section 5:	• The Square Pyramid Numbers.
144		– The 3-Dimensional Sri Yantra.
145		• Conclusion.
146	PART 3B	• An Original and Better Way to Express "e" (the Exponential Function 2.718).
151		• APPENDIX.
151	Section 1:	• Why the Number 9 is Uniquely Inter-Dimensional.
152	Section 2:	• An Advanced Earth-Heart Meditation.
159	Section 3	• Platonic Solid and Phi Spiral Diagrams for General FotoCopy Use.
165		• ACKNOWLEDGEMENTS of GRAPHICS USED and NOTES.
167		• BIBLIOGRAPHY
170		• INDEX.
178		• NUMBER'S INDEX.: The Harmonic Stairway.
181		• A Description of Jain's Books and Video.
188		• The Importance of Base 12, Jain's Petition and Prophesy.

A NEW DIMENSION TO INFORMATION TECHNOLOGY.

DEDICATION:

1) – to the Restoration and Reconstruction of
the **Global Temple of Mathematics**
for the Children of all Nations
that it be a Nucleus that awakens Whole Brain Holographic learning
whose Curriculae instils fun, discovery and
the Supreme Art of Pattern Recognition

2) – also dedicated to the mysterious 9899 whose reciprocal:

$$\frac{1}{9899} = .0001010203050813$$

as an aesthetic emblem for all the devotees of Phi
in this Now Age of "No Devotees As We Are All Masters"
to become a Reciprocal, as it were,
an Ambassador for the Divine Plan / Plane / Planet
to see God / Goddess in All People,
in all Creatures in the Mineral, Crystal,
Elemental (Fire, Earth, Air and Water) Realms, in all Forms in all Dimensions

3) to the Fusion of Pent and Hex

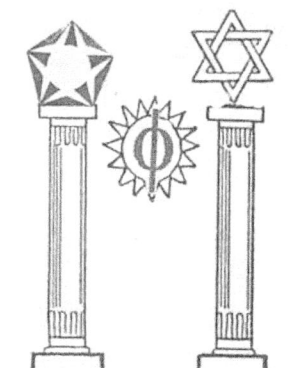

A VISUAL LECTURE ON PHI Φ

including

- **VEDIC MATHEMATICS**
- **MAGIC SQUARES**
- **THE PLATONIC SOLIDS**
- **THE FIGURATE NUMBER SERIES**

JAIN

2003

PART 1A

QUESTION:

HOW CAN YOU DETERMINE THE VALUE OF PHI
(Φ = 1.618033)
"THE GOLDEN MEAN"
USING ONLY
VEDIC MATHEMATICS,
AND NO CALCULATORS,
TO 6 DECIMAL PLACES
?

INTRODUCTION:
for MATHS TEACHERS and PROFESSORS:

Today's lecture is based upon **PHI**, the <u>Holy Grail of Mathematics.</u> It's what all these 20 years of intense mathematical studies has led to; its the millennial mystery of Phi that has kept mathematicians like myself, reincarnating to learn, embrace and embody its transcendent qualities, one of which is "its ability to sort out waves and frequencies to make them coherent, embeddable, recursive, self-organising and **SHAREABLE**" (to quote from Dan Winter a scientist who understands Phi). Imagine electronically controlled traffic lights that stop, start, red, green, its quite primitive, many accidents happen at these intersections as the waves collide; but one day somebody invents the "Roundabout" and the traffic just sorts itself out. To this day I've never seen a car accident on a roundabout. "The solution to any problem is **COMPRESSION**" (quote from Dan Winter). If your computers have too much data, the solution is Compression, Fractal Compression. The Line meeting the Circle, the car approaching the Roundabout becomes the Phi Spiral.

Thus, in the mathematical world, the solution to an archaic curriculum is called Vedic Mathematics and it utilises an ingenious **Sutra** called: **"Sammucaya"** which translates as **"Digital Sums"**, but its really about <u>Digital Compression</u>. (See Chapter 7 called "Phi Decoded, The Secret Formula Revealed" of THE BOOK OF PHI by Jain that uses Digital Sums to locate the Pattern in Phi based on 108 (or 12 x 9) and fitting into or corresponding to the 3-Dimensional Form of the 12 pentagonally-faced **Dodecahedron** and hints at the manner in which D.N.A compresses data).

A simple example of Digital Sums is how to solve MULTIPLICATIONS BY ELEVEN as in 23 x 11. This can be done <u>'Mentally and in One-Line'</u> (as per definition of Vedic Mathematics) by a mere additive compression of the two digits 23. We add the 2 and the 3 to get 5 and place that '5' in between the 2 and the 3 to get the Solution which is quickly and simply 253. Or simply split the digits and insert their sum between them:

$$23 \times 11 = 2 \quad 2+3 \quad 3$$
$$= 2 \quad 5 \quad 3$$
$$= 253$$

Today's lecture will involve some of the other 16 Sutras to determine Φ to 6 DP (Decimal Places) utilising and explaining:
- Quadratics / Calculus
- Negative Numbers

- Square Roots
- Duplex

All the above 4 processes are revolutionary principles that solve complex questions in One-Line of intelligent arithmetic. In fact, one of the steps is finding the Square Root of 5 or ($\sqrt{5}$). No mathematician I know can give you the answer with pen and paper; they still rely on the calculator, yet Vedic Mathematics can arrive at the answer swiftly and in One-Line.

One reason why I am here today is to remind each and everyone of you that, if we continue to teach our children the use of calculators to solve all their problems, the global brain, over the next 20 years or the next generation, will dgenerate or atrophy. The ancient art of Mental Calculation is required urgently to rebuild the Mental Muscle of the shrunken Pineal Gland.

To conclude this introduction, I will give, by diagram, a visual representation of Root 5. Fig 1a is the best geometric picture of $\sqrt{5}$, extracted by the diagonal of a Double Square. (Since $\sqrt{5}$ is implicit in the Φ formula $(1 + \sqrt{5}) \div 2$, the ancients took its 3 Dimensional Form: "The Double Cube" and used that as the sacred measurement of their altars: The Eqyptian Priests used this as well as King Solomon). The other two diagrams (Figs 1b and 1c) illustrate how the **Vesica Piscis** contains and generates all sacred symbols and measurements. It is symbolically the meeting and greeting of the Sphere of God/dess with the Sphere of Consciousness of Man/Woman. Each Sphere's Circumference is touching the Heart or Centre of the Other Sphere).

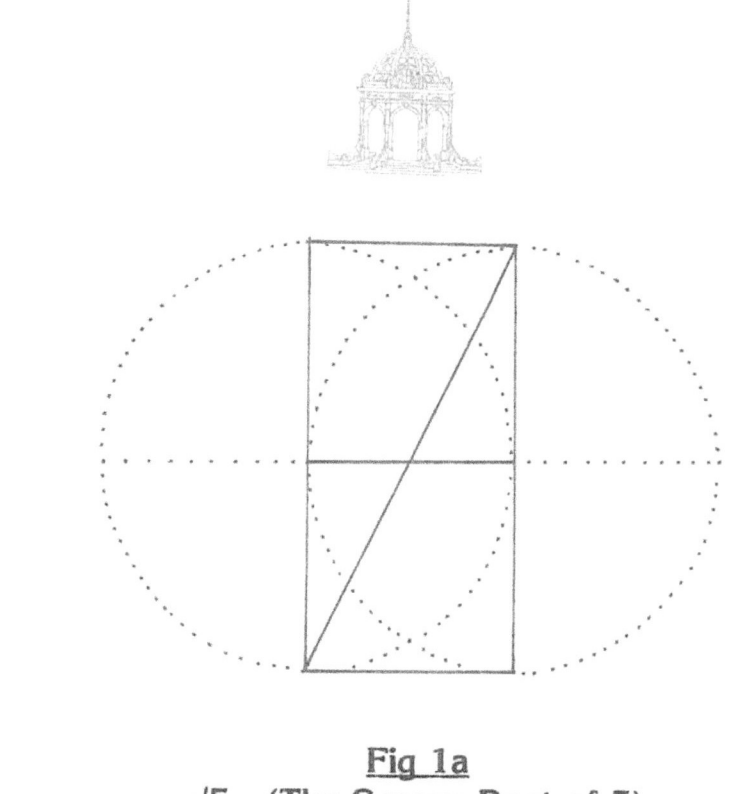

Fig 1a
$\sqrt{5}$ (The Square Root of 5)

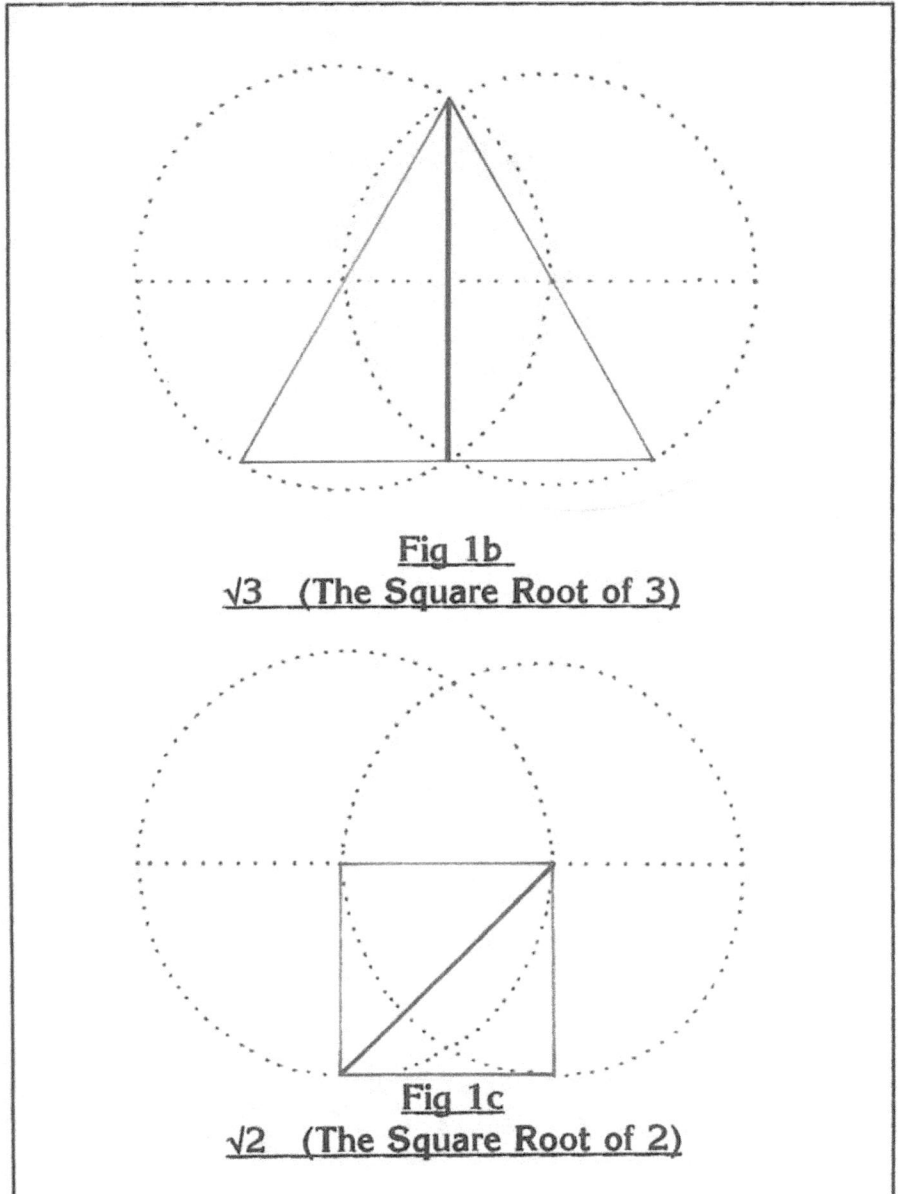

Fig 1b
√3 (The Square Root of 3)

Fig 1c
√2 (The Square Root of 2)

AIM:

Is there a more efficient method for determining the value of:

$$\Phi = \frac{1 + \sqrt{5}}{2}$$

which traditionally requires knowing 2 laborious techniques:

a) how to factorise a quadratic, eg: $x^2 + x - 1 = 0$, using the painful formula:

$$x = \frac{-b \pm \sqrt{b^2 - 4ac}}{2a}$$

(which many students find very difficult to cram and memorize before the dreaded exams!), and

b) how to determine the exact value of √5 to say 6 Decimal Places. This is actually an impossibility in the current western system of mathematics.

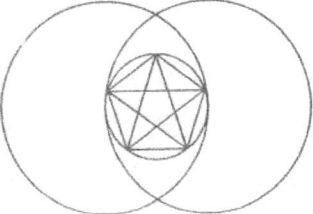

METHOD:

The solution, by traditional methods, for √5, cannot be done, except by calculator. There does exist one archaic method but it would be a sin to attempt showing its difficult procedures.

Whereas, in the 3,000 year old Aryan system, it can be joyfully applied in one line!

And factorising the upcoming quadratic equation ($x^2 + x - 1 = 0$) can also be done in one-line.

HOW TO DETERMINE THE GOLDEN MEAN

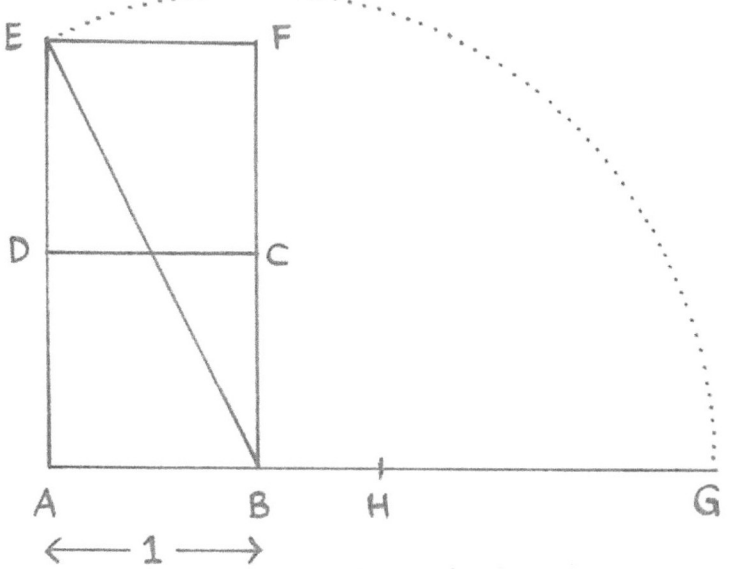

Fig 2
The Double Square Generating The Phi Formula:
(1+√5) Divided by 2

Explaining Fig 2:
- Let A be the centre or zero-point of the intersecting axes.
- Let AB = Unity
- ABCD = Unity Square (ie. 1 unit x 1 unit)
- ABFE = Double Unity's Square
- EB = ABE's hypotenuse = $\sqrt{(1^2+2^2)}$ = $\sqrt{5}$ Diagonal of the Double Square
- EB's arc is traced to G
- H = midpoint of AG = The Golden Mean
- Since BG = BE = $\sqrt{5}$, then
  AH = $(1 + \sqrt{5}) \div 2$
     = Φ = **1.618033**
  BH = .618033 = the Reciprocal of Phi or $1 \div \Phi$
  (or $1/\Phi$) = AK in Fig 9.

If we did not know about the geometry of the Double Square and were asked to determine the value of the Golden Mean "B" of a given line "AH", as shown in Fig 3a (a simplification of Fig 2), such that
$$\frac{BH}{AB} = \frac{AB}{AH} = \Phi$$

Fig 3a

This is the ancient puzzle of The Golden Mean. It is posed in Euclid's books "The Elements" circa 300BC:
THE WHOLE SEGMENT (fir**ST**)
IS TO THE LARGER SEGMENT (seco**ND**)
AS IS THE LARGER SEGMENT
IS TO THE SMALLER SEGMENT (thi**RD**).
ie: What is the relationship between the proportions of the fir**ST** to the seco**ND** to the thi**RD**?
ie: What is this ancient, timeless, universal Phi

ST ND RD ?

Let us call the unknown distance BH = x and add this to our former diagram:

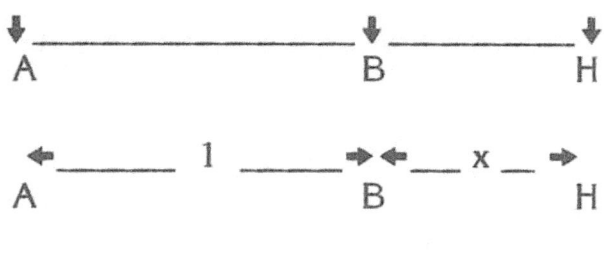

Fig 3b

We know that AB = 1 and BH = x and the whole distance AH must = x + 1. Apply this to the given formula:

BH/AB = AB/AH

i.e: $\frac{x}{1} = \frac{1}{x+1}$

Cross-Multiplying, we have:
$x(x+1) = 1 \times 1$
$x^2 + x = 1$ Bring over the 1 to the other side and therefore reversing the (+1) to a (−1) to arrive at:
$x^2 + x - 1 = 0$

Using Vedic Mathematics to solve this quadratic equation quickly:
We say that the
DIFFERENTIAL = ±√ DISCRIMINANT
$2x + 1 = \pm\sqrt{[(1)^2 - 2.(1).2.(-1)]}$
(where the dots represent the operation of Multiplication)
$= \pm\sqrt{[1--4]}$
$= \pm\sqrt{5}$
$x = \frac{-1 \pm \sqrt{5}}{2}$

We are half way there. We have ascertained that there are two roots:
$x = (-1 + \sqrt{5}) \div 2$ and
$x = (-1 - \sqrt{5}) \div 2$
but we don't know what the value of √5 is yet which is the second half of this article. Until then, here are some more explanatory notes defining

a) how the Differential is the Square Root of the Discriminant which makes obsolete the formula we may already know of as:
$x = \frac{-b \pm \sqrt{(b^2 - 4ac)}}{2a}$

b)   And some notes on <u>Negative Numbers</u> and the concept of <u>Duplexing</u> (which you need to know to understand how to do Square Roots in One-Line) which introduces another Sutra called "<u>Vertically and Crosswise</u>".

In **Vedic Calculus**, we **Differentiate** by multiplying the Power by the Coefficient and putting the answer as the new coefficient. And the new Power is the old Power minus one.

eg.1: $8x^4$ expressed in a "differential way" is
$(4 \times 8)x^{4-1}$ or $32x^3$

eg.2: $x^2 + x - 1$ $= 2x^{2-1} + x^{1-1} - 1^{1-1}$
$= 2x + x^0 - 1^0$
$= 2x + 1$

The **Discriminant** of a Quadratic Expression is the Square of the Coefficient of the middle term minus the product of twice the coefficient of the first (x^2) and twice the last term.

eg.1: $3x^2 + 8x + 5 = 8^2 - 2.(3).2.(5)$
$= 64 - 60$
$= 4$

eg.2: $x^2 + x - 1 = 1^2 - 2.(1).2.(-1)$
$= \sqrt{(1--4)}$
$= \sqrt{5}$

This is useful:
(DIFFERENTIAL = $\pm\sqrt{}$ DISCRIMINANT) as not all quadratics will factorise into whole numbers as in our specific case determining the value of Φ.

[An example of easy factorising in whole numbers is:
Find the factors of the quadratic:
$x^2 - 5x + 6 = 0$
Answer is $(x - 2)(x - 3) = 0$
Therefore, $x = 2$ or $x = 3$]

"ON NEGATIVE NUMBERS"

This article involves the use of what we call **"Negative Numbers"**

On the previous page we were determining the value of $\sqrt{5}$ to ascertain a value for Phi and we came across this enigma: "$1--4$" $= 1 + 4 = 5$?

It was established that:

[the DIFFERENTIAL = $\pm\sqrt{}$ DISCRIMINANT
$2x + 1 = \sqrt{[(1)^2 - (2\times1)(2x-1)]}$
$= \sqrt{[1--4]}$
$= \pm\sqrt{5}$
$x = \dfrac{-1 \pm \sqrt{5}}{2}$]

What does this really mean?
How can 1 minus minus 4 = 5.?
Yet, without this use of Negative Numbers, we could not arrive @ the Golden Mean, nor have created a sophisticated technology, no computers, nor sent rockets into outer space.

Think of a Negative Number as the Magnetic Polarity of a Positive Number.

All Numbers have a Consciousness and a Polarity/Duality. Just like we have Man and Woman,

Night and Day. Originally, it was was recorded in ancient Hindu texts and merchandised accounts with Fibonacci of C12th Italy that if someone owed you money, that <u>debt</u> was written as "–" or negative. This was convenient again to explain the polarity of Abundance (+) versus lack of (–). Negative Numbers are just part of our Reality. We need them. Some scientists visualise +ve (positive) Energy as Centrifugal or outwardly radiating Energy, whilst the Spiral going inwardly is Centripetal or –ve (negative) or feminine or receptive. Understanding the concept of Negative Numbers is understanding not just the reality of "<u>Explosion</u>" but equally and as importantly, the meaning of "<u>Implosion</u>".

(You may want to research ancient Egyptian Mathematics as they had no use of Negative Numbers but rather an Infinity of whole Numbers growing ever bigger, like 1,2,3,4,5,6,7,8,9 etc and corresponding reciprocals or fractions, like 1/2, 1/3, 1/4, 1/5, 1/6, 1/7, 1/8, 1/9 approaching the Infinity of Zero ! A totally different concept and perhaps contradictory to this Article. But it must be considered as we encounter it upon our Path, in search of Mathematical Truth.

THE LAST THING YOU NEED TO KNOW IS:
"DUPLEXING"

As we approach the Vedic One-Line Method of Square Roots, as in √5, we need also to understand the polarity or opposite process of Square Rooting which is really the ability to "Square" numbers. This process is called "**Duplexing**":

eg: To find 23 Squared or 23^2 in One-Line, we need to look at a Pattern, viz:

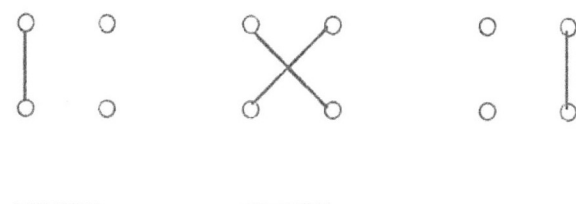

Fig 4a
Sutra: 'Vertically and CrossWise'

This Sutra, known as: **"Vertically and Crosswise"** shows we will have a 3-Digit Answer represented by the 3 horizontal short lines.

The next diagram shows how we get the answer

to 23 Squared: 2 3
                    x
                  2 3

Notice there are 4 digits involved. Let each digit be represented by a small circle and that explains the format seen in Fig 4a. This will help you to understand "CROSS-ADDITION" which we will perform in the centre of Fig 4b:
(Here, the symbol 'x' = Multiplication):

$$= \quad 2 \times 2 \quad (2 \times 3) + (2 \times 3) \quad 3 \times 3$$

$$= \quad 4 \quad\quad\quad 1\underline{2} \quad\quad\quad 9$$

$$= \quad 5 \quad\quad\quad 2 \quad\quad\quad 9$$

= 529

Fig 4b

The DUPLEX concerns the central cross-over " X " which is really a form of cross-addition. To obtain the central digit, we say:
$(2 \times 3) + (2 \times 3) = 12$
We put down the '2' and carry the '1'.
Actually, there is a Short-Cut for quickly determining the Duplex of any Squared 2-Digit Number and that is to merely look at the "23" and double the multiplication of both digits. Thus the Duplex of 23 is:
$D(23) = 2 \times 2 \times 3 = 12$

Another example:
The Duplex of 69 Squared is:
$D(69) = 2 \times 6 \times 9 = 108$

Thus to have the ability to find the Square Roots of Numbers, we also need to know the Duplex of a Single Digit, and the Duplex of Double and Triple digited numbers etc.

The Duplex of a Single Digit is merely its Square. In Fig 4b it is represented by the 3 x 3. Thus the Duplex of 3 Squared is $3 \times 3 = 9$.

Knowing this, we can now solve 23^2 using Duplexes:
$23^2 = D(2) \ / \ D(23) \ / \ D(3)$
$\quad\ = 2^2 \quad / \ 2 \times 2 \times 3 \ / \ 3^2$
$\quad\ = 4 \quad / \quad 12 \quad / \quad 9$ (the "1" gets carried over)
$\quad\ = 529$

What about solving the Squaring of 3 Digit Numbers like 236^2 or 4 Digit Numbers like 2360^2.

Let us look at their Vedic Pattern. Remember that when we use <u>Pattern Recognition</u> it is Feminine Brain Mathematics.

For 3 Digit Numbers being squared, the Pattern to remember is:

Fig 5

For 4 Digit Numbers being squared, the Pattern to remember is:

Fig 6

Regarding Fig 5, looking at the central cross, you can now determine the Duplex (D) of a Triple Digit Number like 236^2. Its short-cut is merely to: "double the product of the outer digits and add the square of the middle digit":

$D(236) = (2 \times 2 \times 6) + (3)^2$
$= 24 + 9$
$= 33$

Regarding Fig 6, looking at the central cross, you can now determine the Duplex (D) of a 4 Digit Number like 2360^2. It's short-cut is merely to: "Double the product of the outer 2 digits + double the product of the inner 2 digits".
$D(2360) = (2 \times 2 \times 0) + (2 \times 3 \times 6)$
$= 0 + 36$
$= 36$

Knowing the Pattern now, you could predict the Pattern for the Duplex of a 5 Digit Number Squared, like 23606^2. It would look like this:

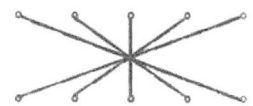

Fig 7

and the subsequent cross-additions is:
"Double the product of the outer 2 digits + double the product of the next 2 inner digits + square the central digit":

$D(23606) = (2 \times 2 \times 6) + (2 \times 3 \times 0) + (6^2)$
$= 24 + 0 + 36$
$= 60$

And just for fun, the Pattern for a 6 Digit Number Squared, like 236067^2 would look like this:

Fig 8

You are ready now to understand how to solve √5, based on the Duplex principle shown as the central cross-additions in Figures 5, 6, 7 and 8.

Our task now is to complete the extraction of the roots of the Golden Mean by determining the approximate value of √5 to say 6 Decimal Places. In Vedic Mathematics, this can be done easily in one line of working but in 'Western' Mathematics there is no known method.

Here is the setting out for the **Square Root of 5** and its immediate answer:

$$\sqrt{5} = 4 \overline{\left) 5. \;_1 0 \;_2 0 \;_4 0 \;_4 0 \;_7 0 \;_{10} 0 \;_{12} 0 \right.}$$
$$\phantom{\sqrt{5} = 4)\;\;} 2. \;\; 2 \;\; 3 \;\; 6 \;\; 0 \;\; 6 \;\; 7$$

The following 7 steps, to determine √5 to 6 Decimal Places, are merely the mental steps required which involve the subtraction of specific Duplexes:

• <u>Step 1</u>: The Square immediately under 5 is 2^2. Place the 2 under the 5 and carry the remainder 1 to make a 10.
This '2' is doubled and becomes our Divisor throughout the whole operation.
• <u>Step 2</u> : Say 4 into 10 is 2, and carry the remainder 2 to make a 20.
• <u>Step 3</u> : After this moment, say 4 into 20, but we have to begin subtracting the Duplexes in their increasing increments ie from 1-Digit, to 2-Digit, to 3-Digit, to 4-Digit etc. (This process of subtracting Duplexes always begins with the digits immediately following the Decimal Point).
Here, we say $= [20 - D(2)] \div 4 = [20 - 4] \div 4$
$= 16 \div 4 \;\; = 4$ rem 0 or 3 rem 4.
(where 'rem' = remainder).

- **Step 4**: = [40 − D(23)] ÷ 4 = [40 − 12] ÷ 4
= 28 ÷ 4 = 7 rem 0 or 6 rem 4.
- **Step 5**: = [40 − D(236)] ÷ 4
= [40 − (24 + 9)] ÷ 4 = 7 ÷ 4 = 1 rem 3.

So far, our answer is 2.2361. The Duplex parts are the numbers after the decimal point. But since D(2361) is greater than 30, we need to change the '1 rem 3' to '0 rem 7'. So that now makes our answer as 2.2360.

- **Step 6**: = [70 − D(2360)] ÷ 4 = [70 − 36] ÷ 4
= 34 ÷ 4 = 8 rem 2 or 7 rem 6 or 6 rem 10.
- **Step 7**: = [100 − D(23606)] ÷ 4
= [100 − 60] ÷ 4 = 40 ÷ 4 = 10 rem 0 or
9 rem 4 or 8 rem 8 or 7 rem 12.

This final 7 gives us √5 to 6 Decimal Places:

$$\sqrt{5} = 2.236067$$

[Interestingly, **Phi** be expressed only in "5"s, eg:

$$\phi = (\sqrt{5})^{.5} \times .5 + .5$$

which means, Phi = root 5 to the power of point 5 (or a half) multiplied by point 5 plus point 5].

EXTRACTING THE 2 ROOTS OF THE GOLDEN MEAN.

Since $\phi = \dfrac{-1 \pm \sqrt{5}}{2}$

then the two possible roots are:
1): [−1 + √5] ÷ 2 and
2): [−1 − √5] ÷ 2.

Substituting the value of √5 = 2.236067 we have:

1) = [−1 + 2.236067] ÷ 2
= 1.236067 ÷ 2
= **.618033**

Technically, .618033 is the reciprocal of Phi (ϕ = 1.618033) and is expressed as a fraction: 1/ϕ. This is geometrically shown as the placement of 'K' in Fig 9.

A poetic way of describing this mystical point 'K' is by quoting the ancient Pythagorean aphorism:
Unity in BiUne Trinity
or
Unity TriVided.

2) = [−1 − 2.236067] ÷ 2
= −3.236067 ÷ 2
= **−1.618033**
= − ϕ (See the placement of 'M' in Fig 9).

We can conclude that the quadratic equation
$x^2 + x - 1 = 0$
has 2 expressions or 2 roots of PHI:
= +.618033
= **reciprocal of Phi or $1/\phi$**
and the other nebulous root being:
$-\phi$

These 2 roots can be extended onto the same original diagram of Fig 2 and is now called Fig 9.

Fig 9 introduces the Realm of Negative Numbers where one of the roots of Phi is located $(-\phi)$ which is shown as point M.

Now study Fig 9 (which is an extension of Fig 2) and observe the 2 possible roots of $x^2 + x - 1 = 0$ as points K and M, the 2 Golden Mean Solutions, one in the real world (K) and the other in its polar imaginary world (M).

It suggests that the enigma of Phi has its basis as an inversion of itself (the reciprocal of Phi, or point K) and a basis in the Other Worlds (symbolized by minus Phi or negative Phi or point M). This explains why Phi is the Implosive, InterDimensional Mathematics of Infinity that bridges the Galactic Worlds with the Atomic Worlds, and since we Humans are created totally in the Phi God

Proportions then we are that Bridge between worlds symbolized by this phenomena of the reciprocal and negative roots. It means we need to learn the art of Inversion, like the Hanged Man, we need to learn the art of surfing between the Dimensions. That's what Phi says, its the perfect sorting of waves so they don't collide, its knowing how to turn ourselves inside-out and outside-in.

Referring to Fig 9 again, we understand that Point A is Zero or Point 0 or Center, so let us look at the relationship of the 2 roots in 2 ways:
1): We can say that MA to AK is really a proportion of $-\phi : 1/\phi$ or $-1.618 : .618$ and the real distance between them is MK
= $-1.618 + .618$
= -1.

What does this '-1' mean?

Remember our initial postulate was to begin with a Unity Square, ie a Square whose 4 sides are 1 unit each. This is symbolic of our constant reference to <u>UNITY CONSCIOUSNESS</u>. That's why, in Fig 9, it is referred to as AB (literally, "the Father"). Thus if Point B = the projection of Unity (or +1) then its mirror image or projection/reflection into the Other World is at Point N (or -1). We symbolically started off on our journey at +1 and after a Lifetime we arrived at -1. It means that Phi lends us the ability to

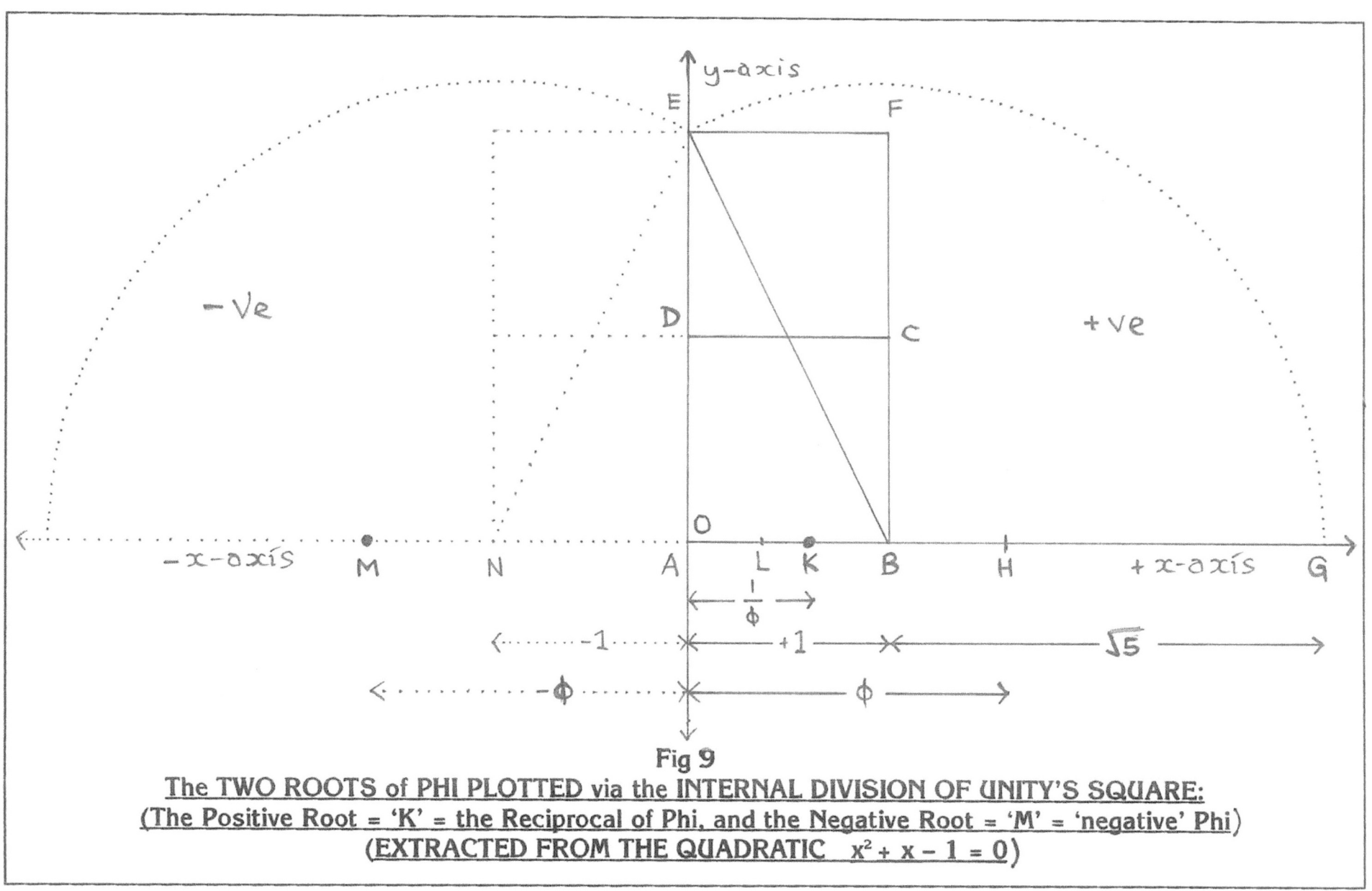

Fig 9
The TWO ROOTS of PHI PLOTTED via the INTERNAL DIVISION OF UNITY'S SQUARE:
(The Positive Root = 'K' = the Reciprocal of Phi, and the Negative Root = 'M' = 'negative' Phi)
(EXTRACTED FROM THE QUADRATIC $x^2 + x - 1 = 0$)

navigate between worlds, we have to know how to steer through the Sun symbolised by the Zero Point at Point A, where the Axes of Spin cross.

Before discussing the other relationship of the roots, and whilst we are analysing this phenomena of Negative One (–1), let us state one of the basic definitions of Phi, and that is:

PHI ALLOWS ITS WAVES TO ADD AND MULTIPLY

Let us therefore check that the addition and multiplication of these two extracted roots are the same. This is a good way to determine if the roots are correct:

$(-\phi + 1/\phi) = -1.618 + .618 = -1$ and
$(-\phi \times 1/\phi) = -1.618 \times .618 = -1$

Therefore we can conclude that these are the correct roots and they do indeed add and multiply to the same wavelength. This is the ability to TIME TRAVEL! This is the symbolism of the negative or minus sign, that it is self-similar to our world or dimension or axes, but that it is parrallel but in another World that has another axis or axes of spin. The Phi Ratio allows us to move between Worlds and Time and Dimensions without collision.

2): The other possible relationship of the 2 roots is to measure the distance between the graphed Points M and K linearly. Even though AM and AK represent different worlds of polarity (–ve and +ve) we could get out our earthly ruler and say that the distance between the two roots of M and K is:
$1.618 + .618 = 2.236067$ which is of course
$= \sqrt{5}$ (the original Diagonal of the Double Square that we began this exercise with!).

Thus the final answer was already pre-knowledged or indicated in the original question:
THE QUESTION IS THE ANSWER. Check for yourself in Fig 9 that the distance of EB = MK = $\sqrt{5}$.

To conclude Part 1 of this article I would like to give a bit more information about why $\sqrt{5}$ is important and where else it appears, before leading onto Part 2 which locates the golden mean division within Unity's Square symbolised by AB and becomes an exercise for the interested student to practise their newly learnt skills in Vedic Mathematics.

Imagine a Cube whose dimensions are 1 x 1 x 1. Well in the Theosophical School of Thought, representations of the Universe and Atomic worlds are seen as nestings of the 5 Platonic Solids, known as the Greater and Lesser Maze where geometric objects are placed one inside the other like Chinese dolls: The Octahedron nested within The Star Tetrahedron or Stella Octangula nested within The Cube nested within The Icosahedron nested with The Dodecahedron, allowing your D.N.A to ignite and share memory from The Micro and Macro Worlds. If you were to measure this nesting of geometric models, in relationship to the original 1x1x1 Cube you will find a predomination of √5 and Phi as the faces of one shape are stellated to morph into another shape or another Atomic Structure. This is called Alchemy.

Keep in mind also that the dual roots of $-\phi$ and $1/\phi$ (symbolised by M and K) are not really measureable as an exact number, but rather the reality of <u>PHI is a Pure Principle, a Function of Infinity</u> whose unending decimal is 1.618033 and seemingly travels ad infinitum without any known recursion or detectable pattern in its cryptic decimal form. Well this is what has been taught for thousands of years but if we return our attention to simple whole numbers, like the Fibonacci Sequence that generates the Phi ratio, then I can declare that I have located or discovered an inner pulse or rhythm or pattern that does repeat. Just like on the previous pages we began with a tangible model based on Unity, the 1 x 1 x 1 Cube, our Reference Point, our Beginning and our End, we were able to locate this entity called Phi, but not touch it. Similarly, we can teach any Child about the famous Arithmetic Sequence called the Fibonacci Numbers that are whole numbers and their subsequent divisions generating this mystical non-recurring decimal of Phi.

The beauty of being able to arrive at 1.618 yourself, from whole numbers, like dividing preceding terms of the Reduced Fibonacci Numbers or like the content of this Article is supreme and easy to comprehend, whereas a Scientist may be able to tell you that he/she can generate Phi to ten thousand decimal places on their sophisticated computer according to some unheard of and cryptic formula involving logarithmic functions which you don't know anything about; and yet these same scientists would caution any apprentice to be wary of anything that you can't prove or replicate or observe or touch, suggesting that it may be part of Borg-Hive-Mind-Control-Techno-Hex-Based-Consciousness. Any alien coming into this Sphere or Planet could quickly ascertain the Consicousness of this Planet by inspecting our Educational Curriculum. If they saw the way we do Mathematics, they would smirk, and perhaps move on. Our planet is lacking the grace of the dimensions of the Golden Mean. It is time to restore this once glorified and ancient system of Divine Intelligence.

It is also of interest that √5 is part of the formula that generates Phi [(1 + √5) ÷ 2]. This is our first clue to the Pattern in Phi. If we square √5 we get 5. The raising of a number or a frequency or a vibration ($x \Rightarrow x^2 \Rightarrow x^3 \Rightarrow x^4$ etc) is a shift into the next dimension. That's why the Number 5 predominates largely in the literature on Phi depicted as the Pentacle:

This very symbol gives us a clue as to where the Pattern or Pulse or hiccup or recursion is within The Golden Mean: (Refer now to THE BOOK OF PHI, Chapter 6: THE PHI CODE REVEALED by Jain, 2002). Remember, in the Question is the Answer. The Pentacle reveals the Code of 108:

Fig 10
The 2-Dimensional (2-D) Pentacle

The external angle of the Pentacle (shown in Fig 10) is **108°** and this is the mystical number and secret of The Golden Mean. It is based on the 12 Pairs of **Reduced Fibonacci Numbers** that all sum to 9: (12 x 9 = 108). The act of digitally reducing numbers to single digits is the <u>Mathematics of Compression</u>.

There are also many other correspondences to the number 108 like the 108 angles of the Lesser Maze that nested the Platonic Solids.

A perfect polyhedron

Fig 11
The 3-D Dodecahedron: 12 Pentacled Faces

The 3-Dimensional or holographic form of the 2-Dimensional Pentacle is **The Dodecahedron** (Fig 11) which conveniently packs and accepts the 12 Pairs of 9 data for D.N.A.'s genetic evolution, whilst incorporating the √5 measurement.

Now that you realise the importance of Shapes and 3-Dimensional models, and how they pack themselves self-similarly, you can easily now visualise that if you had 7 Cubes (One of the Five Platonic Solids) on your table and had packed or glued 6 Cubes around the 6 faces of a central Cube you would have what is dubbed a 4th-Dimensional Cube or **Tesseract** (Fig 12). It's really a system of Double Cubes whose measurements are all based on √5 !

Similarly, if you had 13 Dodecahedrons on your table and packed or glued 12 around a central Dodecahedron you would, in effect, have a 4th Dimensional Pentacle.

Shape Stores Memory and bridges the Dimensions, and most effectively when the lengths are based on PHI.

Below is Claude Bragdon's diagram:

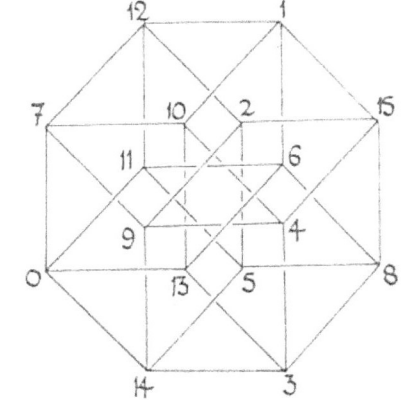

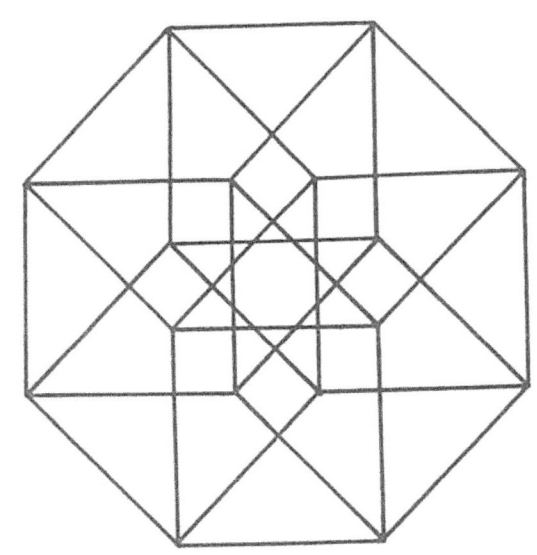

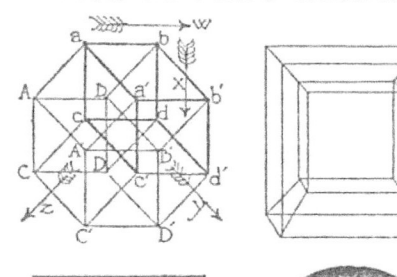

TWO PROJECTIONS OF THE HYPER-CUBE OR TESSERACT, AND THEIR TRANSLATION INTO ORNAMENT.

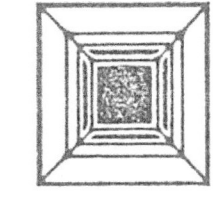

Fig 12
4th-Dimensional (4-D) Cube or Tesseract
and the Magic Tesseract with Magic Sums of 30

A GEOMETRIC SOLUTION

Up to this point, we have just shown the ALGEBRAIC SOLUTION FOR THE INTERNAL DIVISION OF UNITY. (The next pages will shown the Algebraic Solution for the External Division of Unity's Square. For your general interest, I will give you a GEOMETRIC solution for the Internal Division of Unity (AB) symbolised by Point K = .618033 as shown in Fig 13 and first found in print in Euclid's books, of which all our modern mathematics school texts and curriculum are based on.

GIVEN THE LINE AB (which = 1)
FIND ITS GOLDEN MEAN DIVISION (K)
SUCH THAT
AK / AB : KB / AK ?

Fig 13 (which is also an extension of Figs 2 and 9) shows how the given line AB = 1 can be internally divided into the Golden Mean @ point K.

Follow this Recipe or Algorithm:
Create the square on AB to form ABCD. Let IB = half of AB. (Triangle) $\triangle$ ABI is right-angled or 90°. Join AI as the hypotenuse of Triangle ABI. Place compass on point I and form the arc IB onto line AI @ J. Place compass on point A and arc AJ onto the original line AB @ K. Since AB = 1 then AK = $1/\phi$ = .618 and KB = $1/\phi^2$ = .382

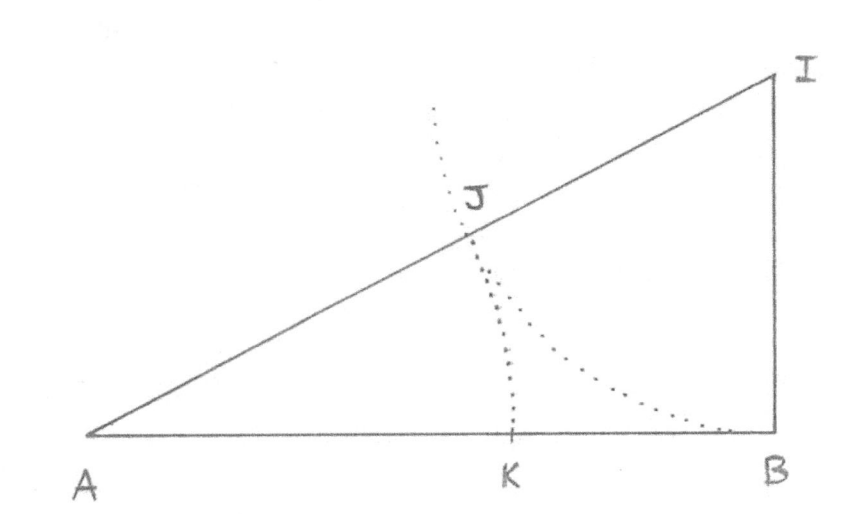

Fig 13
Using A Compass To Determine
The Divine Proportion

Now that you have read and understood Part 1 of this article and you know how to evaluate Phi via clever Vedic Mathematical techniques without the use of the dreaded calculator, you must congratulate yourself for graduating out of Factory / Hive-Mind / Borg / Half-Truth Mathematics that has kept humanity limited to quite a degree for millennia.

You have just evaluated one root as the reciprocal of Phi. Proceed now to evaluate the other root of Phi.

PART 1 B

Determining the Dual

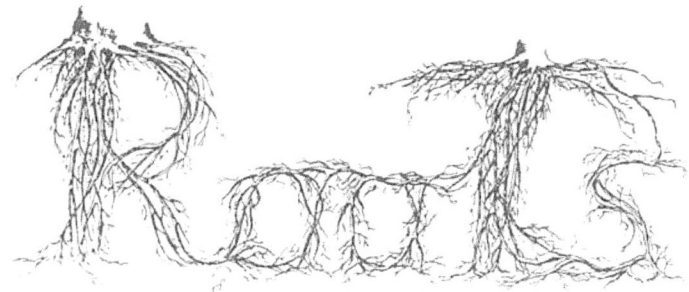

of

PHI

The "Extreme Mean Ratio"

via

External Division

and

Internal Division

of Unity's Square

This chapter is based on the millennially old problem of finding the Golden Section of a Straight Line AB. It was first documented more than 23 centuries ago in **Euclid's Book 2, section 11**. (Pythagoras was teaching @ 500 BC and Euclid, his student, @300 BC).

Don't allow the following mathematics to disturb you. If it is over your head, just gloss over it and come back to it later, to just appreciate that the ancients were in search for a mystical division of Unity, symbolised by the 1 x 1 Square or 1 x 1 x 1 Cube and that the Solution involved a Pair of Numbers, one that has "x" is Greater than 1, representing the Macro or Universe, and one that has "x" Less than 1, representing the Micro or Atomic World.

For this reason, it was termed: "The Extreme Mean Ratio or Division where we seek 2 Dual Roots or 2 answers or 2 magical gateways achieved by cleverly allowing:

- "x", the UNKNOWN FACTOR to be Greater than Unity, and therefore called THE EXTERNAL DIVISION OF UNITY, and
- "x", is Less than Unity, and therefore called THE INTERNAL DIVISION OF UNITY.

Observe the graphic of the Tube Torus Doughnut in Fig 35 having Dual Openings (which occurs in the first day of your creation geometries, one opening becomes your mouth and the other your anus).

Fig 14 a
TORUS HAS +VE and -VE OPENINGS

You will observe that Duals appear everywhere in Nature, like the 2 poles of a magnet, like the Duals of the Platonic Solids, shown here as Fig 14b.

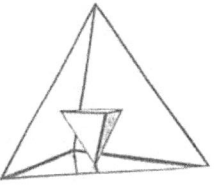 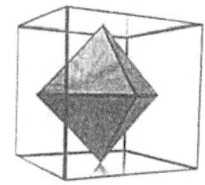 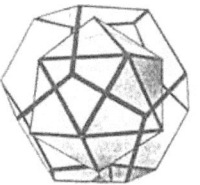

**FIG 14 b
THE DUALS OF THE PLATONIC SOLIDS**

Magic Squares also have a sophisticated pairing system (see Fig 49a showing The Lo-Shu or Magic Square of 3 x 3) where all the Paired Numbers around the central digit 5 have a sum of 10
eg: 1+9 =10,   7+3 =10,
6+4 =10,   2+8 =10

6	1	8
7	5	3
2	9	4

This is really symbolic of another form of 'Extreme Mean Division' because we are saying that the magic centre of 5, the Mean, is a result of the midpoint of the sum of the two extreme numbers 1 and 9; and all Pairs of 10 revolve about the centre cell. Having the pairs aligned, thus creates a matrix where all the rows, columns and diagonals can now have and share the same harmonic number vibration of 15.

DETERMINING THE ROOTS OF PHI USING EXTERNAL DIVISION OF UNITY'S SQUARE

So far, we just determined the Internal Division of Unity's Square (AB), a number less than 1, to be .6180339887 known as the reciprocal of Phi (or $1/\phi$) or ($\phi - 1$). It is also known as the <u>GOLDEN RATIO CONJUGATE</u> and sometimes also called the <u>SILVER RATIO</u> expressed numerically as ($\sqrt{5} - 1$) ÷ 2. This is called the Positive Root and its Negative Root was determined to be ($-\phi$) as shown in Fig 9 (Points K and M respectively). Glancing at Fig 2 shows how we began to achieve the Internal Division of Unity's Square by calling BH = "x" the Unknown Factor, where BH or "x" is Less than 1. Even though we are dividing the line we are really expressing a Trinity and for this reason it is called the triVision of UNITY or UNITY triVided.

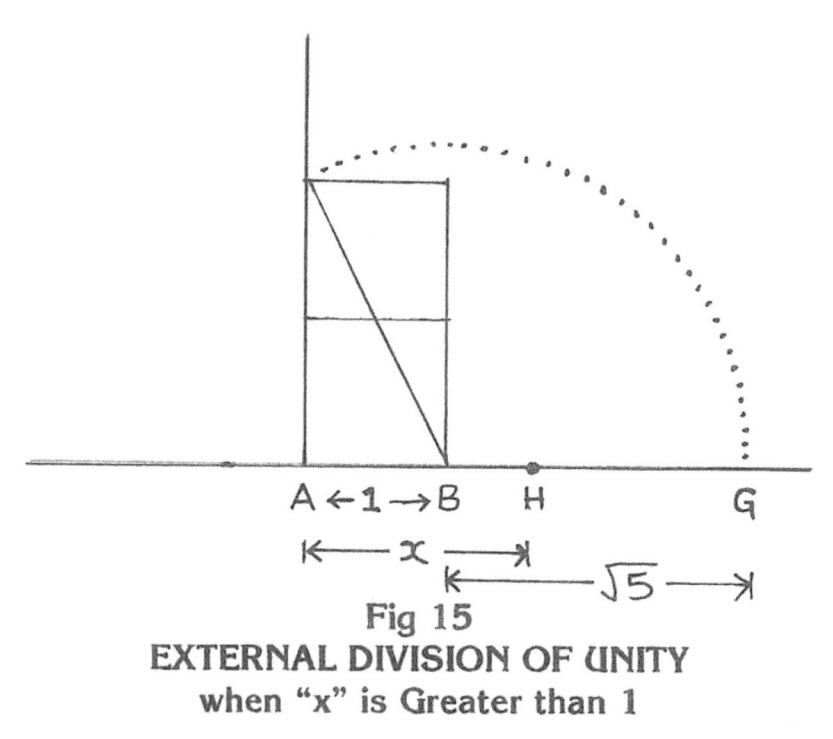

Fig 15
EXTERNAL DIVISION OF UNITY
when "x" is Greater than 1

To achieve the final answer, (point H which is the PHI point of 1.618033) we need to make "x" the unknown factor, to be Greater than 1, to achieve the External Division of Unity, as shown in Fig 15. It will have a Positive (+ve) Root of $+\phi$ (Point H) and a Negative (–ve) Root of $-1/\phi$ (Point Q).

Applying the Vedic Sutra of Proportionality (called ANURUPYENA) we compare:

$$\frac{AB}{AH} : \frac{BH}{AB} \quad \text{ie.} \quad \frac{1}{\text{"x"}} = \frac{\text{"x"}-1}{1}$$

Cross-Multiplying, we get $x^2 - x = 1$

becomes $$x^2 - x - 1 = 0$$

To solve, or determine the dual roots of this Quadratic Equation, using Vedic Mathematics, we say:

<u>The Differential</u> = <u>The Square Root ($\pm \sqrt{}$) of the Discriminant</u>

$$2 \cdot \text{"x"} - 1 = \pm \sqrt{[(-1)^2 - 2.(1).2(-1)^2]}$$
$$= \pm \sqrt{(1+4)}$$
$$\text{"x"} = \frac{1 \pm \sqrt{5}}{2}$$

This means that for the External Division of Unity there are 2 Roots: a) one in the +ve realm and b) one in the −ve realm.

a) +ve "x" $= \dfrac{1 + \sqrt{5}}{2} = \dfrac{1 + 2.23606}{2}$

$= 1.618033$

$= \phi$

b) −ve "x" $= \dfrac{1 - \sqrt{5}}{2} = \dfrac{1 - 2.23606}{2}$

$= -.618033 \quad = -\dfrac{1}{\phi}$

OBSERVATION: 1

$x^2 - x - 1 = 0$ can also be written as:

$\phi^2 - \phi^1 - \phi^0 = 0$ NOTICE THE INDICES or the POWERS which are **2, 1, 0** which expresses Phi as a laddering of the Dimensions, an orderly descension of Powers or Dimensions involving a Trinity of Powers of an unknown entity GOD.

SUMMARY: CHART OF THE 2 QUADRATIC EQUATIONS AND THEIR DUAL ROOTS

For "x" Greater than 1 EXTERNAL DIVISION		For "x" Less than 1 INTERNAL DIVISION	
$x^2 - x - 1 = 0$		$x^2 + x - 1 = 0$	
POSITIVE ROOT	NEGATIVE ROOT	POSITIVE ROOT	NEGATIVE ROOT
$+\phi$	$-\dfrac{1}{\phi}$	$+\dfrac{1}{\phi}$	$-\phi$
(+1.618)	(−1.618)	(+.618)	(−1.618)

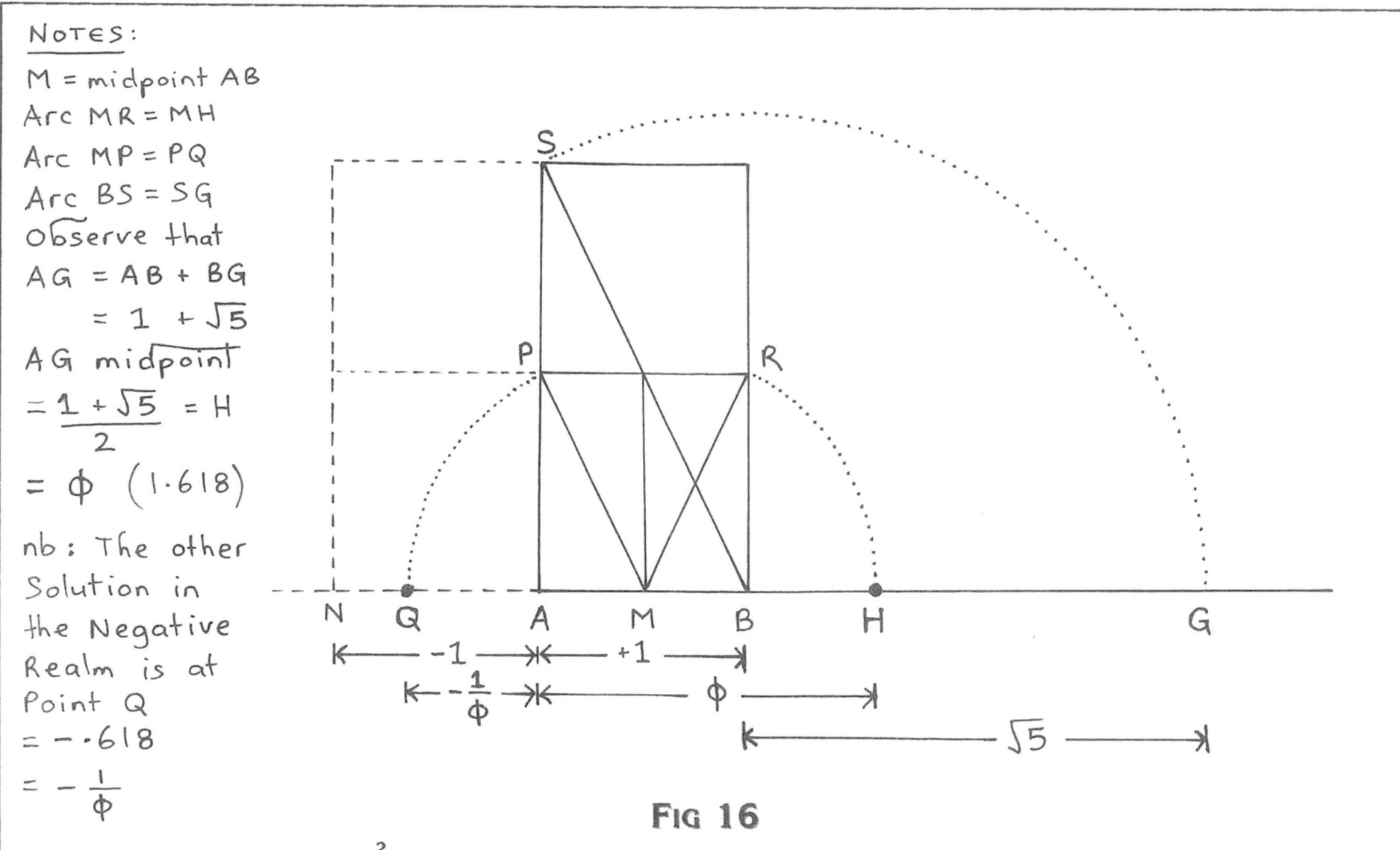

FIG 16

THE DUAL ROOTS OF $x^2 - x - 1 = 0$. **The External Division of Unity.** AKA $\phi^2 - \phi^1 - \phi^0 = 0$

QUESTION:
What is the best symbol that graphically captures all this mathematics, for someone that doesn't understand all this algebra, how could you totally simplify this mathematics so they 'get it'?

ANSWER:
The ROSE.
The Rose to me is the ultimate symbol for the Phi Mystery as the supreme packing of its many petals capturing the concept that the Golden Mean isn't just this number 1.618033 etc but rather a principle of many all inclusive cascading harmonics expressed by the Fibonacci Sequence.
The Rose also implies a nesting of these frequencies, and an embedding of the same, a self-similarity wrapped into a divine twirl, with a great smell.

OBSERVATION: 2

Phi has this unique property of being the only known number in the Universe, when lessened or diminished by 1 it becomes its own RECIPROCAL!

ie. $\phi - 1 = \dfrac{1}{\phi}$

Cross-Multiplying: $\phi^2 - \phi = 1$

$$\phi^2 - \phi - 1 = 0$$

thus the solutions are ϕ (1.618) and $1/\phi$ (.618)

OBSERVATION: 3

Another unique property is that (the roots of) PHI knows how to both "ADD AND MULTIPLY" itself:

Looking at only the +ve roots in the above Chart for $x^2 - x - 1 = 0$

Adding ϕ and $-1/\phi$ = 1.618 - .618 = 1
Multiplying them = $\phi \times -1/\phi$ = -1

QUESTIONS AND ANSWERS

QUESTION: If Phi relates to Root 5 (or the Diagonal of the Double Cube), how can a simple person apply this in their daily life?

ANSWER: In regards to Feng Shui, the Art of Aesthetic Placement, or Vastu Shastra as it is known in India, it means that whatever is the Height of your Body, use that distance again above your Head, to determine the Height of your House's Ceiling.

You see, your vertical body height is equal to the distance of your horizontal and outstretched arms! So in effect, as you pivot on the spot, you are creating a cubic vessel. The Cube is your bio-etheric container on Earth. By doubling your Cube, you create the Golden Mean and therefore establish an implosive link to interDimensional access and to the Bliss Experience. High ceilings make you feel good. Geometry opens a Door.

Phi, Vedic Maths, Platonic Solids, Magic Squares and the Tetrahedral Nos.

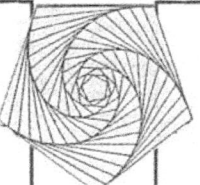

YOU ARE PHI
the Living Temple of
Mathematics.
The Roots
of your Formulae
are in This World
and The Other
extending through
the Souls of your Feet
and Crystal Pineal Glands
connecting you
to all the Kingdoms,
Above and Below,
Inside-Out
and
Within.

AN IDEA:

The quadratic $x^2 + x - 1 = 0$ is really
$x^2 + x = 1$
(which I suspect is very similar to the
Fractal Formula,
the Mathematics of Infinity).

What this is really saying is that:
SOMETHING, let us call it 'x' plus
this SOMETHING SQUARED (x^2)
equals UNITY:

$$x^2 + x = 1$$

This then is the
<u>Mathematics of Unity Consciousness</u>,
hinting at the ability of our Time-Travelling
ance stors known as the ELS or The Elders Race
who knew how To Turn,
how to enter and exit the Dimensional Worlds
through what they called a
<u>90° Phase Shift of Consciousness</u>,
for only the Phi Spiral / Torus Doughnut /

Smoke-Ring Shape / Imploded Sphere
has the ability to Turn itself
Inside-Out and Outside-In

(The Fractal Formula below shows how data (C)
is repeatedly looped backed or ITERATED
into the Equation,
growing towards an infinite resolution.
The 'C' means that the output of one operation
= the input of the next one.
But like above formula,
there is this relationship between The Something 'x'
and The Something Squared 'x^2',
or a 90 Degree Turn
which is the adding to the next dimension):

$$X \approx X^2 + C$$

NB: To go from x to x^2
is really a step or travelling into the next dimension.
That's what it means to square or cube, is really to
add another axis of spin to the reality, like adding a
z axis to the existing x and y axes to form a cube,
to go from a Square (x^2) to a Cube (x^3)

and even to a HyperCube (x^4).
This then becomes a
Jacob's Ladder of Dimensions represented by the
Powers of 'x' or Axes of Spin.
The beauty of this Fractal Formula
and therefore the Phi Formula,
is that not only does this iteration continually grows
larger and larger, it can also move
infinitely smaller and smaller, somehow looping or
connecting the Macrocosm or Galaxies with
the Microcosm or Atomic Worlds.

This subject really now comes under a new
heading perhaps best defined as
INTER-DIMENSIONAL MATHEMATICS.
It is the true meaning of this book entitled:
<u>IN THE NEXT DIMENSION</u>
which was inspired by my American publisher
(<u>Sacred Science Institute</u>)
who publishes books on Vedic Mathematics and
thought it was my specific task to link the two
seemingly different subjects of
Sacred Geometry / Magic Squares / Phi
to the 16 Universal Sutras
that constitute the ancient body of knowledge
called Vedic Mathematics.

THE 24 REDUCED FIBONACCI DIGITS OF THE PHI CODE INSERTED IN SEQUENTIAL ORDER AROUND A MANDALA OF 24 RAYS SUCH THAT ALL OPPOSING PAIRS HAVE SUMS OF

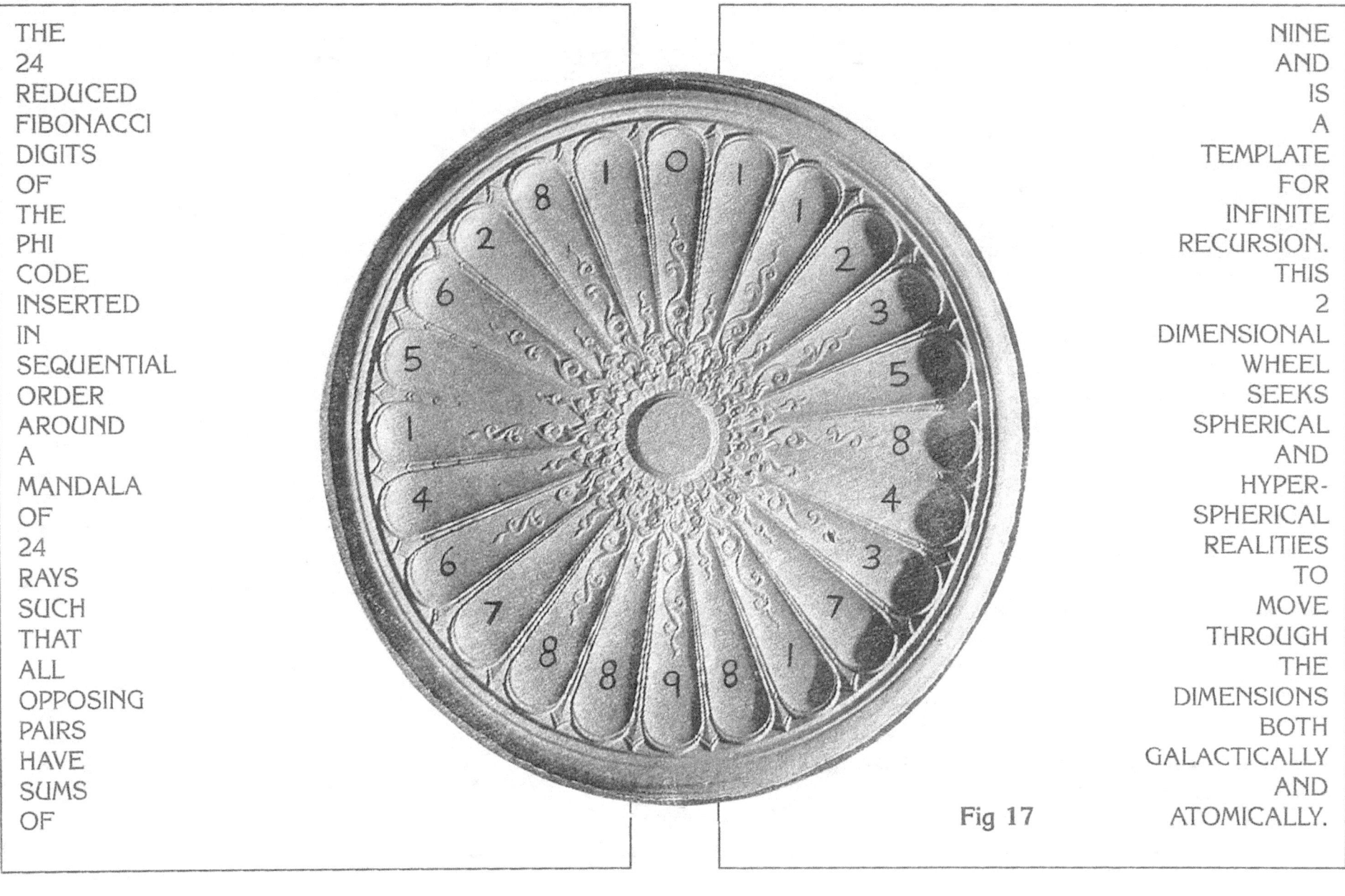

Fig 17

NINE AND IS A TEMPLATE FOR INFINITE RECURSION. THIS 2 DIMENSIONAL WHEEL SEEKS SPHERICAL AND HYPER-SPHERICAL REALITIES TO MOVE THROUGH THE DIMENSIONS BOTH GALACTICALLY AND ATOMICALLY.

PART 2

MORE NOTES ON THE
PHI CODE

incorporating
Duals of the Archimedean Solids, 4th Dimensional Tetrahedrons (Pentatopes) and HyperCubes

---

The 12 pairs of 9, as a configuration, are more stable than the string of 24 digits (that marvellously recurs infinitely, each set summing 108).

So rather than thinking of a
1)- <u>Star-Tetrahedron</u> (inter-Digitated Platonic Solid Tetrahedron) having 24 faces and 24 edges, and
2)- <u>CubOcatahedron</u> (Archimedean Solid, one of the 13) having 24 edges,
it is more economical and compact and practical to envision a 3-D model that has 12 faces (relating to the 12 Pairs of 9), and that must be:
the **Dodecahedron**.

For this to be true, each of the 12 pentagonal faces must have a quality or polarity of 9ness (12 Pairs of 9).

There is another 3-D model that fits this paradigm, not a Platonic Solid that has equal lengths and equal edges and all vertices having the same shapes, but an Archimedean Solid, somewhat similar but having different polygons meeting the vertices. There are 13 such Archimedean Solids, like the traditional soccer ball shapes of pent and hex, and the one that is of relevance here is the <u>**Dual of the Truncated Tetrahedron**</u> (3,6,6 configuration) known as the <u>**Triakis Tetrahedron**</u> (3,3,3 configuration but having isosceles and not equilateral triangles. Shown below in Fig 18 are 2 facets of the Truncated Tetrahedron + the net for the Triakis Tetrahedron:

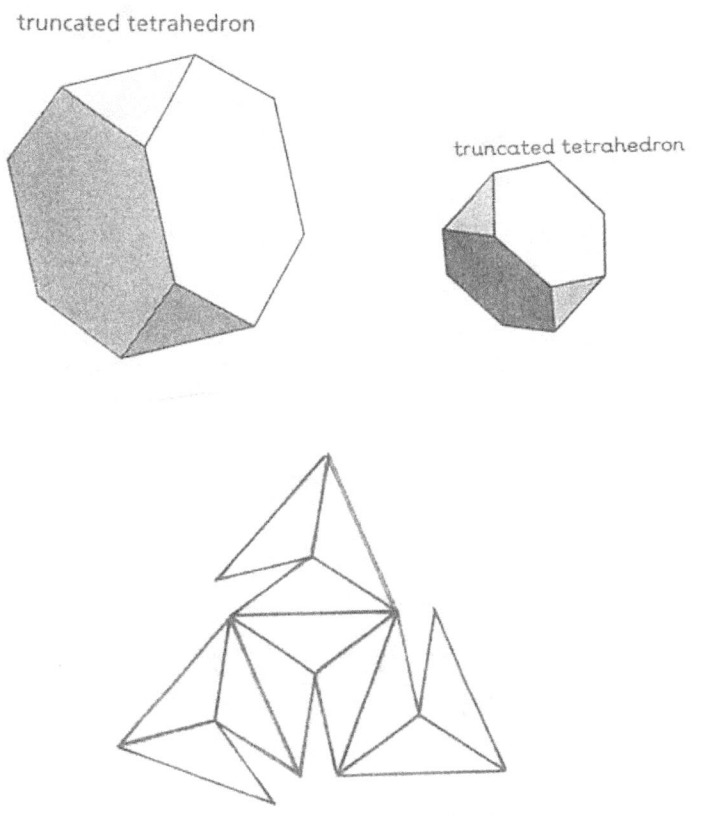

Fig 18
Net for the **TRIAKIS TETRAHEDRON +
2 VIEWS OF THE TRUNCATED TETRAHEDRON**

Whilst we are exploring all our possibilities of Shape Management let us summarise that the Phi Code Pattern is based on
24 recurring reduced Fibonacci Numbers,
12 Pairs of 9 and
6 Pairs of 9 in the Alternate Number Palindromic Sequence of the 24-Code that knows how to internally reverse the sequence in pairs of 6 !

So what 3-D model or crystal knows how to incorporate this quality of 6, 12, 24.

The only crystal shape I know of is the CubOcatahedral crystals of an important ore of silver known as **Argentite** (Ag_2S) or Silver Sulphide, a dark, lead-grey sectile, crystalline, massive mineral (from the latin word for silver, 'argentum').

The CubOctahedron satisfies this 6, 12, 24 code by having 24 edges, 12 faces, but the 12 faces are really 6 Squares ☐ and 6 Triangles △. (This defines it as an Archimedean Solid as each vertex does not have the same polygon, but rather, in this case, a triangle and a square. If you were to hold Buckminster Fuller's famous model of this known as the Jitterbug (Fig 19) which knows how to compress itself into all the 5 known Platonic Solids, you would be able to visually see that the 6 Triangles form 3 opposing pairs of Triangles, but each vertically upward Tri△ngle is opposed diametrically by a vertically downward tri▽ngle to form ✡ when viewed from a distance, looking toward centre. These opposing triangles △ ▽ suggest the palindromic or reversed sequence of the Phi-Code.

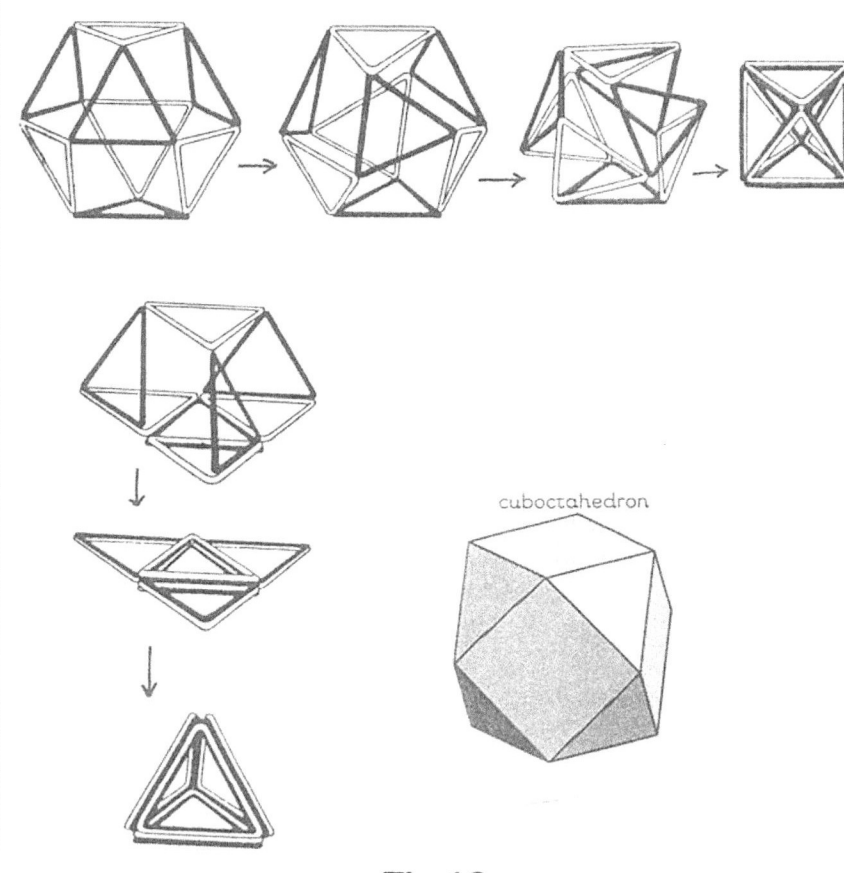

Fig 19
B. FULLER'S CUBOCTAHEDRON "JITTERBUG"

This research material is purely inspirational as well as mathematical and scientific. These models suggested are not fixed in stone and other ideas are welcome. You may add that the 12 Pairs of **9** are an important part of this Phi Code and if asked to relate this 9ness to the 12 Faces of Dodecahedron I would ask, "How can the pentagonal faces of the Dodecahedron express a quality, vibe or polarity of 9ness?" I would therefore begin to show the divisions of the Pentagram within the Pentagon as shown in Fig 20 below:

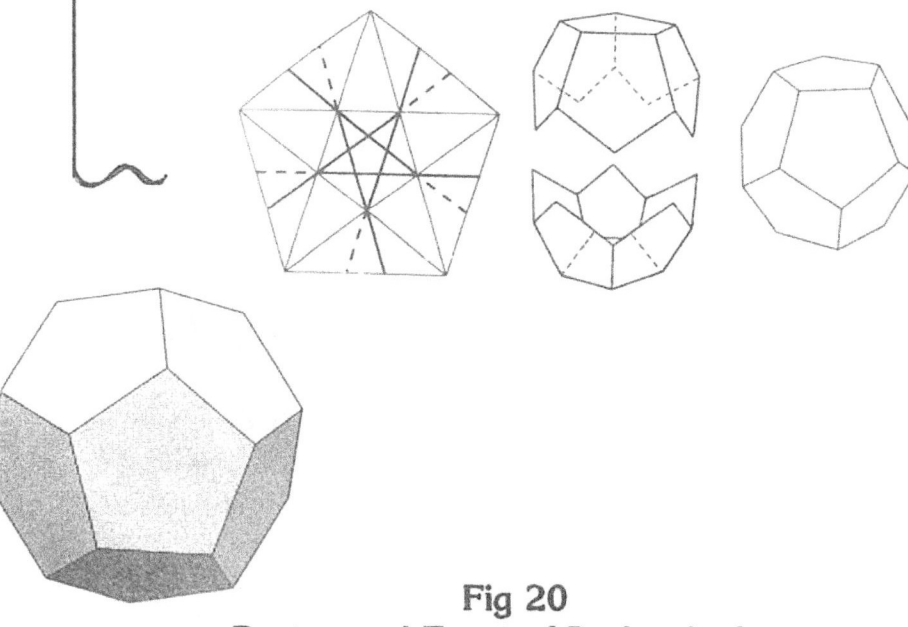

Fig 20
Pentagonal Facet of Dodecahedron

In figure 22 below you probably think you are looking at a 2-D (2-Dimensional or planar) figure. You are, but it is also more than that.

When examining the Phi Code (of 24 Reduced Fibonacci Digits) we cannot limit ourselves to just the 3rd Dimensional (3-D) view. We must go beyond. Para-Gatay. We must ask ourselves:

"WHAT SHAPE IS ALSO VERTEX INSCRIPTABLE IN HYPERSPHERES?"

Such an answer would be to make the clear analogy that in the 4th Dimension we have the HyperCube or Tesseract (Fig 12) having <u>24 Squares</u> (+16 Vertices +32 Edges + 8 Cubes all in one model).

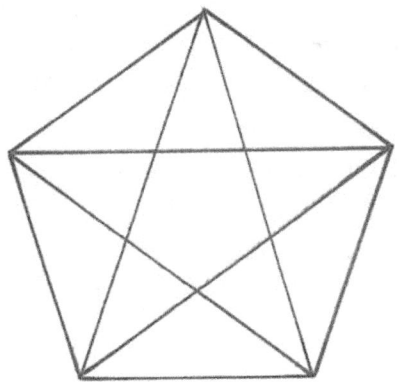

**Fig 22
THE PENTATOPE (4-D TETRAHEDRON)**

Observing the centre as void and regarding the 10 triangles so formed, we can place or superimpose the 10 digits of our Hindu numerical system into this yantra or mystic diagram (0, 1, 2, 3, 4, 5, 6, 7, 8, 9). Our original enquiry was to paint a picture of 9ness into the canvas of the dodecahedron. A possible solution to this is shown below in Fig 21 where all opposing pairs of numbers sum to 9
eg: $0 + 9 = 9$, $1 + 8 = 9$, $2 + 7 = 9$ etc

The Golden Section in a Star.

$AB = .382$
$BY = .618$
$AY = 1.000$
$XY = 1.618$

**Fig 21
Pentacle-Pentagon with Opposing Pairs of 9**

Similarly, regarding Fig 20, we must ask ourselves:

"THE SHADOW OF WHAT 4TH DIMENSIONAL OBJECT IS PENTAGONAL ?"

It is the **Pentatope**, the simplest figure in 4-D. Known as the representation of the 4-D analog of the the Platonic Solid Tetrahedron; it is also known as The Five Cell or **The 5-Cell** as it has 5 Vertices.

Suppose you have on your table 5 tetrahedrons. Around a central tetrahedron, called ABCD in 3-D, glue the remaining 4 tetrahedrons to form the 4-D Tetrahedron. The Dual of this Pentatope is ItSelf, having 5 lots of 3-D Facets, 10 Ridges or Faces, 10 Edges and 5 Vertices ! ie it is Self-Dual: If you were to connect the centres of the 10 Faces from inside and have a peek you would see the same shape.

Another way to grok this important 4-D image is to think of the 3-D Tetrahedron's 4 Vertices as ABCD. Now visualise a chosen point E that magically traverses along the 4th Dimension through the centre of ABCD so that EA = EB = EC = ED = AB.

'Imagination is more important than knowledge'
... Albert Einstein

Whilst we are on the subject of visualizing geometric polygons associated to the Phi Code of the 24 Reduced Fibonacci Digits, let us have a peek at a 24-sided polyhedron (Fig 23) with cubic symmetry but is an idealized re-entrant unit cell produced by symmetrical collapse.

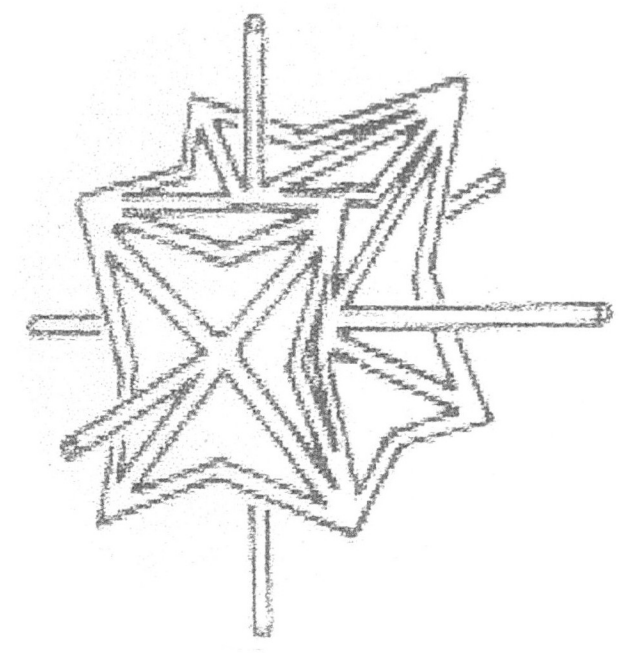

Fig 23
Idealised Imploded 24-Sided Polygon

24 SPHERES IN 4-D

I'd like to alert you about some of the great dedicated mathematical researchers on this planet, like J. H. Conway and N. J. A. Sloane who co-authored the book:
"Sphere Packings, Lattices, and Groups". There is a great web-site for figurate number-theory enthusiasts by Sloane called the:
"On-Line Encyclopedia of Integer Sequences", whose handle is www.research.att.com/cgi-bin/access.cgi/njas/sequences/A.cgi?Anum-001116 where you can find thousands of universal sequences. In this case we are investigating the Phenomenon of the 24 Phi Cycle and are researching the number of spheres that can touch or be packed around a central sphere in multi-dimensions. The following Sequence is from this web-site:

0, 2, 6, 12, <u>24</u>, 40, 72, 126, 240, 272...

It's called the:
"MAXIMAL KISSING NUMBER OF n-DIMENSIONAL LATTICE".
Looking at this Sequence tells us, at a glance, that:
- In 2-D, there are 6 Circles around a central Circle, (as shown in Fig 24a,
- In 3-D, there are 12 Spheres around a central Sphere (as shown in Fig 24b), and
- In 4-D, there are <u>24</u> (Hyper)Spheres around a central (Hyper)Sphere.

There's that magic omnipresent number again !

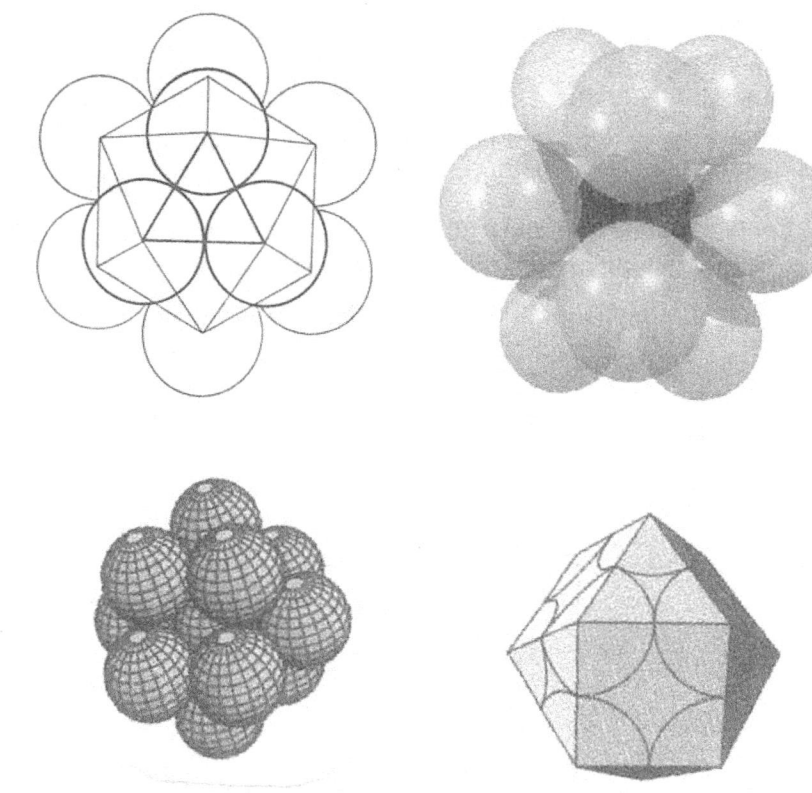

Figs 24a and 24b
Circle/Spheres IN THE NEXT DIMENSION

It is for this reason, and many others that I have titled this book:

IN THE NEXT DIMENSION

When you can start thinking "In The Next Dimension" the secrets of Sacred Geometry will come to you. Simply, if you see a Triangle, think of it as a Tetrahedron; if you see a Square, it is the shadow of a Cube. Also, the typical Hexagon is another 2-D view/shadow of the Cube. Similarly, a five-pointed pentacle in 3-D is really the Dodecahedron. A Circle is The Sphere. Can you thus imagine what a 2-D Swaztika would look like in 3-D? Fig 25 uses the convenient cubic container to give you some insight.

**Fig 25
3-D VIEW OF THE SWAZTIKA**

In such manner, the 24 Reduced Fibonacci Numbers are not just a linear Sequence. It prefers to form Multi-Dimensional links or models or relationships or realities. The ones that survive must be fractal, literally a branch of the tree, a twig of the branch such that the smallest unit is identical in Shape to the largest unit.

That's why the Double Square geometry and maths shown in this book, that give the value of Phi as $(1 \pm \sqrt{5}) \div 2$ having dual roots with $\sqrt{5}$ gappings or distances or voltages, is really another non-linear sequence, and is holographically, not just the Double Cube but 6 sets of these. Think of a Cube with 6 faces and glue 6 cubes onto these 6 faces to get into the next dimension. Now visualise a smaller fractal Cube within the Larger Cube and you have the makings of the HyperCube or Tessereact (Fig 12) in 4-D that knows how to keep extending itself, scale invariantly, by self-similar holographic tesselations, like Sphere-Packings. This means that the original Double Cube determined a sacred ratio called $(1 \pm \sqrt{5}) \div 2$ and this thus became the sacred template to keep on Cube-Packing in 3-D, 4-D, 5-D, 6-D etc (see Fig 26), not only in the Macro world (+ve Phi) but also in the Micro world (-ve Phi). 'Scale Invariant' means that the size does not matter, only the ratio or proportion of Shapes matter. Thus the perfect solution to wave-

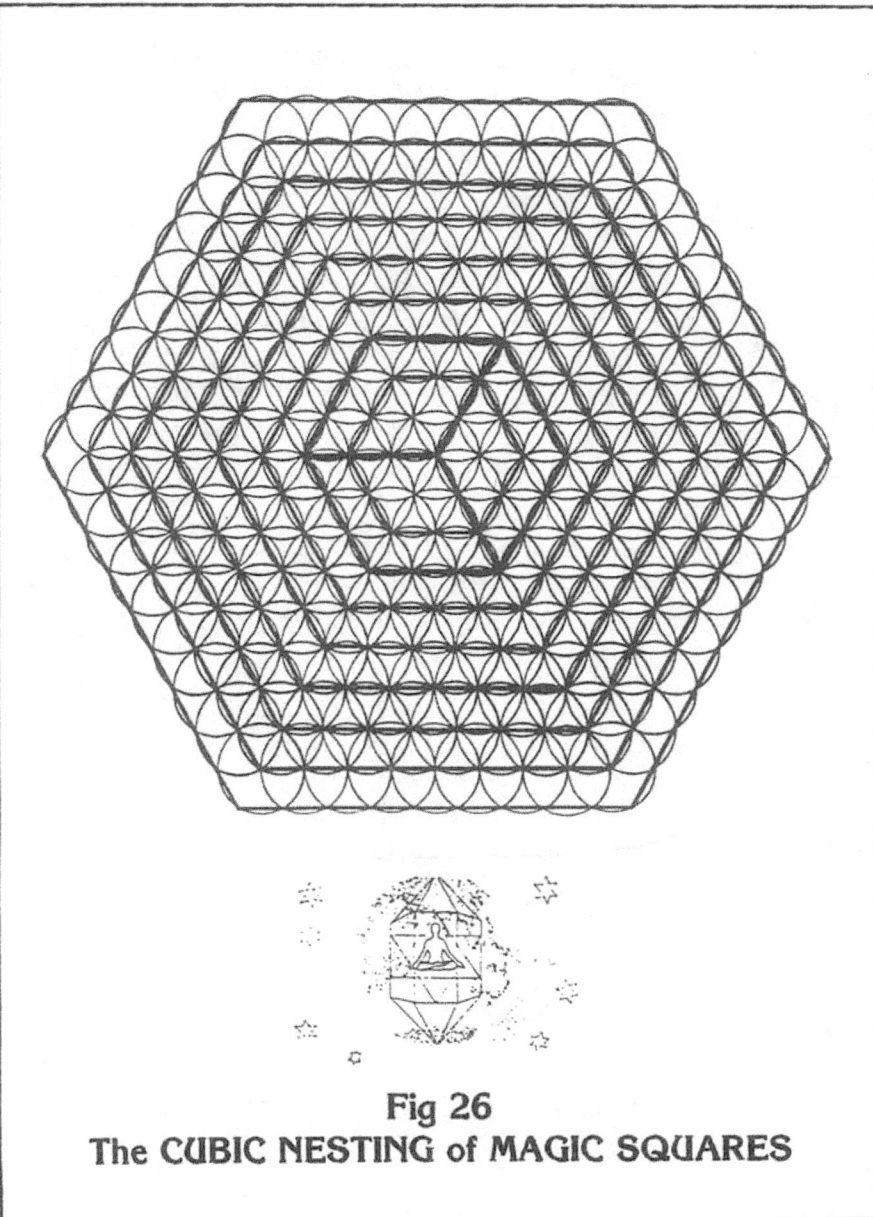

Fig 26
The CUBIC NESTING of MAGIC SQUARES

jamming or chaos is the Phi fractal phenomenon. The Technology of the Future will be based on Shape. Imagine operating computers not via keypads but via Shape. In fact, the problem with our technology is that it is based on Sixness or Hex (German word for 'Witch'). 6 is based on the Cube which is a great container but not if you want to move through the Dimensions. The Cube is great as our Departure Point from which we can Time Travel and safely come back Home again. Its highly efficient in close hexagonal packing as used in graphite inside of lead pencils, (Fig 27 below), which is not really lead, but

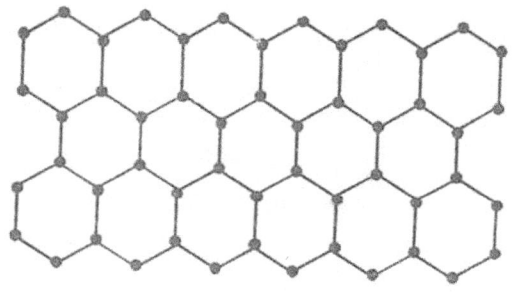

Fig 27
Hexagonal Structure of Graphite

carbon that also comprises Diamond. Diamonds though, are virtually indestructible as it has tetrahedral architecture compared to graphite crystals that grow in hexagonal sheets, explaining why the 'lead' slides off easily onto paper as we write.

Fig 28
HEX / STAR-TETRAHEDRAL COMPUTER CHIP CIRCUIT

HEX VERSUS PENT

There is currently a Star Wars going on with 6 versus 5 or Hex versus Pent.

6 represents the Technology which is based on Silicon Chip circuitry (Fig 28) which is based on the Star-TetraHedron whose vertices join to form a Cube. Dan Winter calls this "IN - <u>CUBE</u> - ATION" as it is all 90 degree angles (forming Squares). There's no real movement, compared to water which crystallises at 60 degree angles (forming Triangles).

The bee-hive mind, also hex, is descriptive of the compass whose radius of 1 unit remains unchanged and fits around its Circle 6 times, exactly. Its obedient, its group mind consciousness, the workers do the work, the nectar is produced. But what happens when the radius can go beyond and within its parameters, and still produce nectar. This is what PENT does as symbolised by the PHI RATIO of the PENTACLE and the DODECAHEDRON.

Pent wants to transform the mechanical technology, that is <u>outside of us</u>, into something more organic and biologically based on Nature's ratio, to something that is <u>inside of us</u>. Nature has ensured that **<u>all our living proteins are 5-sided architectures</u>**! The implied Star-Wars is one of converting computer-based Hex technologies to Pent-computers based on Living Proteins. Don't you think its a bit primitive that when you are sitting in front of a computer there is this hidden electron gun firing incessantly at your third-eye. And some of the people that built these monstrosities have an intelligence that know only how to dump their effluent into the sea! A serious re-education needs to take place, one that begins in the home, how we learn to recylce our rubbish, from compost to effluence, a re-education in the schools that reintroduces the lost 16 Sutras of Vedic Mathematics into the curriculum. You see, if we continue to relie on calculators (Hex) we are giving our power to this technology. Vedic Mathematics allows the student to become quicker and better than the calculator, it stimulates the Third Eye, improves memory and confidence and is very empowering. Same with healing. Do you choose to go to a doctor who will give you a strange chemical pill to 'fix it', or do you make time to cleanse and heal yourself. Do you choose free, non-polluting water-powered cars or the ugly and destructive vehicles we use today?

In this world of dualities, we need to be in both worlds of Hex and Pent comfortably, so a union of the two shapes is required.

Fig 29 shows how the Vesica Piscis forms the Square, Rhombus and Hexagon, in one diagram, used here to illustrate 'the Above and the Below'.
Fig 30 demonstrates the fusion of Hex and Pent.

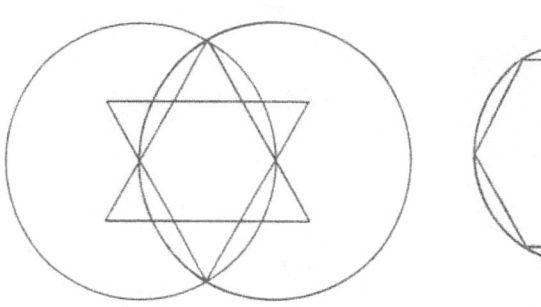
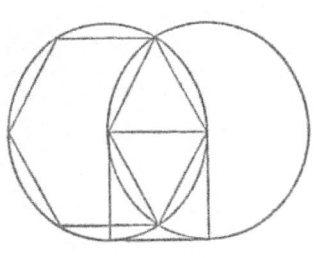

Fig 29
SQUARE, RHOMBUS AND HEXAGON FROM VESICA PISCIS

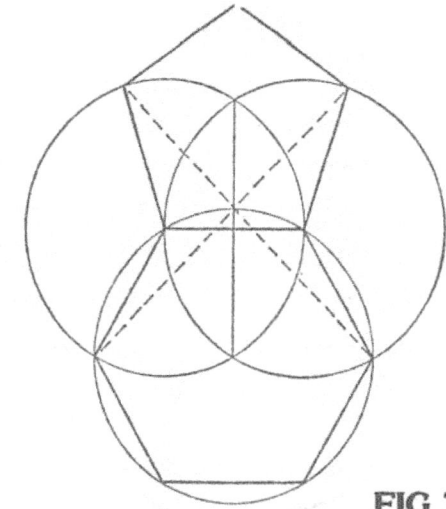

FIG 30
PENTAGON AND HEXAGON CONSTRUCTED FROM VESICA PISCIS

Esoteric geometry was originally passed on and revealed to initiated teachers and students of certain mystery schools or guilds. The challenge of the ancient temple geometers was to create the commonly known pentagon without varying the gap of your compass. We know that the creation of the hexagon is a simple task, but the creation of the pentagon involved more steps. But to create a diagram that fused both Hex and Pent together required more wit (Fig 30). The problem was solved by a famous painter Dürer and published in his "Course In The Art Of Measurement With Compasses and Ruler" (and was reproduced in "The Painter's Secret Geometry" by C. Bouleau). Thus was achieved the union of Hexagon with Pentagon, via the medium of the Vesica Piscis (one sphere's outer edge passing through another and touching its centre).

The ancients were very interested in ascribing numerological correspondences to alphabetical letters creating Words of Power. This was the science of **Gematria** which also investigated geometrical relationships to shapes within shapes and arrived at Sacred Names. A good example is given by John Michell in "City Of Revelation" that says, what if I constructed a Pentagon whose sides were all 1 unit in length and I inspected the relationship of this shape nesting exactly within a Square, and compared the two. When I measure the Square Area of the surrounding Square I determine that it is 2.368 square units. I then look up the dictionary of word values in the ancient Greek Gematria and realise that the name of "Jesus" adds up to 888 and "Christ" adds up to 1480 and when combined they have a sum of 2368. This corresponds to the square area of the the square. (Fig 31). Notice the dotted line which divides the Pentagon into the Phi Ratio. The above rectangle has a square area of .888 (=Jesus) and the lower rectangle has a square area of 1.480 (=Christ).

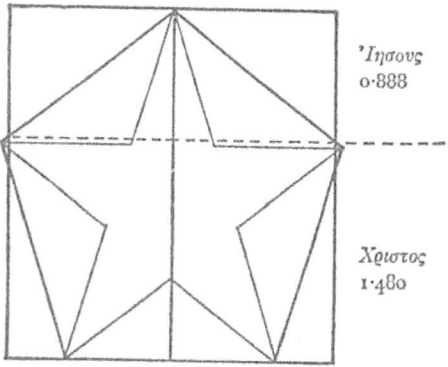

Fig 31
SIDE OF PENTAGON = 1 UNIT AREA OF SQUARE = **2.368**
2368 = JESUS CHRIST (THE GEMATRIA OF THE PENTACLE).

The symbolism of the hexagram and pentagram relates to the dual human condition: the spiritual and the physical selves. This is mirrored within our living D.NA molecules that reveal the same constituent shapes carried within the genetic code. If a chemist was to inspect these nucleic acid/proteins he/she would see two dimensional platelets of two fused pairs of hexagons and pentagons (shown below in Fig 32).

What is interesting here is the mathematics of the rectangle linking the hexagons. Its not just any rectangle. It is the **Golden Phi Rectangle**!

Remember I said before that we need to look at shapes in their true 3-D form. To get a true picture of the D.N.A molecule, as revealed to the world by Doctors Watson and Crick in the early 1950s, we need to look at Fig 33 with 3-D eyes. Imagine if you

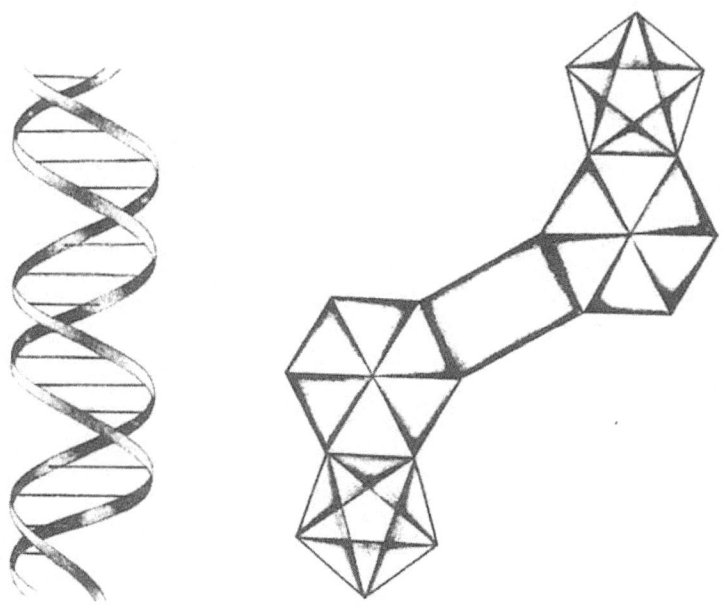

Fig 32a
FUSION OF HEX AND PENT IN THE D.N.A SPIRAL MOLECULE

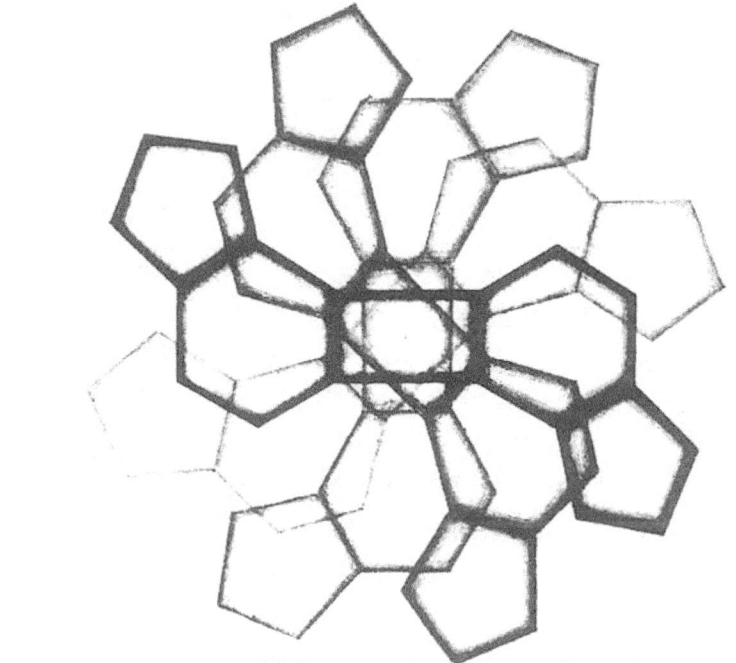

Fig 32b
3-D SPIRAL VIEW OF D.N.A. MOLECULE, TOP DOWN VIEW

were walking up a spiral staircase then you looked down. This is how the rising and twisting rectangles would appear from a top view looking down or a bottom view looking up. This geometric mandala is an essential part of our genetic makeup. Deep inspection or meditation upon such mystical and scientific diagrams would take you to the very core of your being. Fig 34 shows some polygonal views going to centre.

CURVES OF PURSUIT

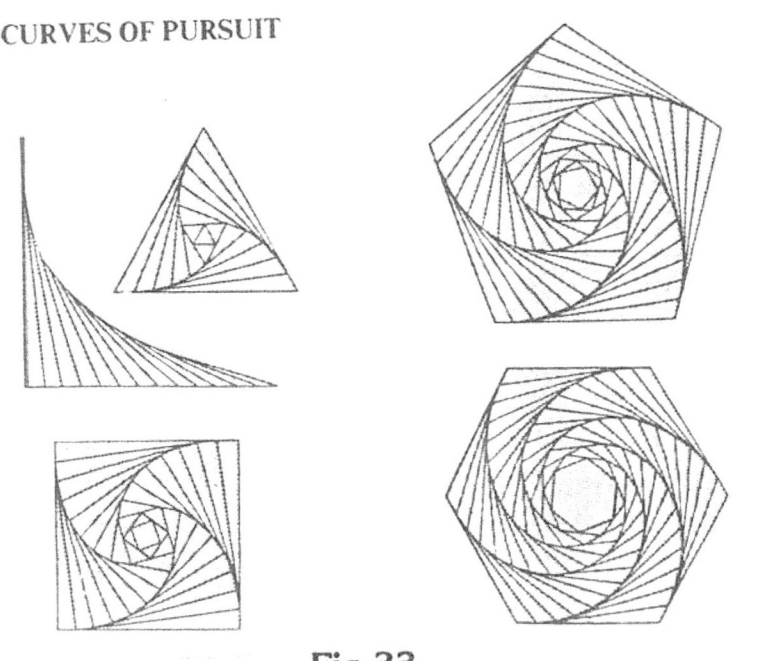

Fig 33
Imploding Pentagon is D.N.A.'s Biological Choice

One of the essential themes of this book is the ability to understand and therefore climb the Dimensional ladder. One of the best graphics that illustrates this is an important sequence of shapes that climbs the Dimensions. If we begin with the familiar Square in 2-Dimensions, we understand that the Cube is its 3-Dimensional counterpart and that the HyperCube is its 4-th Dimensional expression. The diagram below shows the raising of the Square's Consciousness into its Higher Dimensional Forms and will be known here as the 5th, 6th and 7th HyperDimensional Cubes. Meditate upon them.

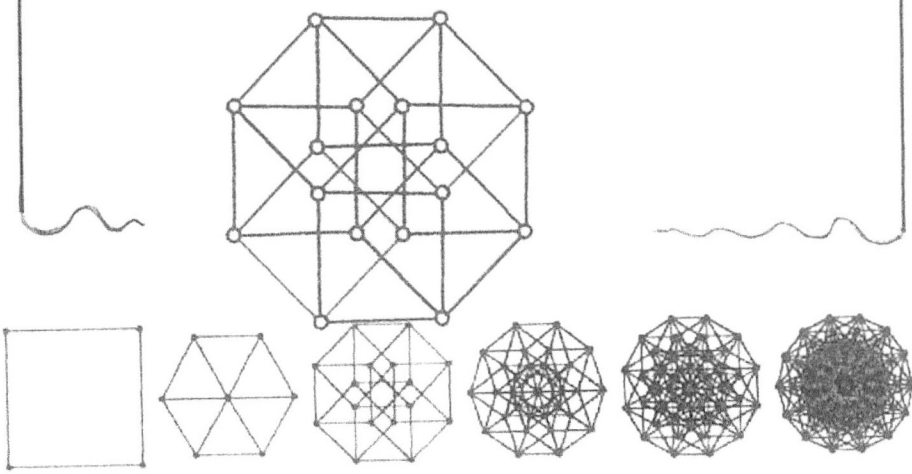

Fig 34
The Sequence of Dimensional Shifts from the Square to the 7th Dimensional HyperCube

PHI and the SUTRAS

THE MAIN VEDIC MATHS SUTRA AT WORK HERE IS:

ANURUPYENA (PROPORTIONATELY)

Remember how we divided the line AB internally, using the double square whose diagonal is √5 and located the Golden Mean or Extreme Division or Divine Proportion at point K (see figs 2, 9 and 15). Shown here again, very simply:

```
|---------------|-----------|
A               K           B
```

We determined that the most aesthetically pleasing ratio or division of the Line AB was at point K such that :

Shiva's dance of the creation and destruction of the world.
(His Nine Consorts, like Shakti, Kali etc represent the 9 Moods of the Divine Feminine Archetype and relate to the 9 Numbers of the Magic Square of 3 x 3).

AK / AB : KB / AK

You can understand that Phi (at point K) is really a proportion between two lengths or 2 numbers. In Sanskrit, the word for 'Proportionately' is ARUPENYENA and is considered as one of the Sub Sutras in Vedic Mathematics. Before I continue this article on Phi and the Sutras, let me give you a few examples how we can use this sub-sutra.

Vedic Mathematics, for the last 3,000 years, uses only 10 digits which are from 0 to 9. When these ancient Hindu scholars considered a number like 11 or 21, they did not think of it in the way we currently do. We think of 21 as a large number, whereas the Hindus looked at 21 as a relationship between the 2 and the 1. Thus if they wanted to square 21 they would do it very quickly and simply: Square the 1, Square the 2 and insert between the two digits "the double of the product of the 2 and the 1":

21 Squared = 2 Squared / 2 x (2x1) / 1 Squared
$\qquad$ = 4 $\qquad$ / 4 $\qquad$ / 1
$\qquad$ = 441

To cube a number, the Hindus could see the predicted answer by application of Right Brain / Feminine / Visuals. Cubic Number solutions were encrypted in the ancient Jaina mathematician Halayudha's Triangle (3rd century BC), also known as Pascal's Triangle. It is shown in Fig 34 as an intelligent cascade of adding neighbouring numbers and writing their sum inbetween and below, a simple process of mere addition:

To solve the cubing of numbers, you merely had to remember a visual pattern that is found in the third row of the Triangle: the numbers 1 3 3 1.

$$\begin{array}{c} 1 \\ 1\ 2\ 1 \\ 1\ 3\ 3\ 1 \\ 1\ 4\ 6\ 4\ 1 \\ 1\ 5\ 10\ 10\ 5\ 1 \\ 1\ 6\ 15\ 20\ 15\ 6\ 1 \end{array}$$

Fig 34
Halayudha's / Pascal's Triangle

These 4 numbers were actually the co-efficients of the binomial expansion $(a + b)^3$.

For a number like 21, a=20 and b=1 and this is substituted into the cubic equation:

$(a + b)^3 = \underline{1}a^3 + \underline{3}a^2b + \underline{3}ab^2 + \underline{1}b^3$

$\qquad = 20^3 + 3 \times 20^2 \times 1 + 3 \times 20 \times 1^2 + 1^3$

$\qquad = 8,000 + 1200 + 60 + 1$

$\qquad = 9,261$

That's the conventional way, and the long way of doing it. Vedic Maths does it mentally utilising the 1,3,3,1, coefficients. This is how they would set it out. The cube of the first digit is put down, ie $1^3=1$

We now employ our knowledge of Anurupyena, the relationship of 2 is to 1 (2 : 1) regarding the number 21 and extend this doubling of the last digit cubed until we have the format of:
$(a + b)^3 = a^3 + \underline{3}a^2b + \underline{3}ab^2 + b^3$

       8   4   2   1

and then double the two central digits and then add them on (as this fulfills the 1,3,3,1 relationship):

= 8   4   2   1
          8   4
─────────────
= 8 ₁2 6 1

= 9,261.

(nb: the "1" of the "12" above is carried over to the left to make the "8" a "9").

- Understanding this, we can mentally calculate 11 cubed (11^3) which requires a 1 : 1 proportion

= 1 1 1 1
=   2 2
─────────────
= 1 3 3 1

- 32 Cubed (32^3) requires a 3 : 2 proportion after cubing the last digit:

= 27 18 12 8
=   36 24
─────────────
= 27 ₅4 ₃6 8
= 32,768

What I am establishing here
is that Vedic Mathematics does not seek
to nullify Western Mathematics,
but rather to enhance it.
We are interested in a zen-type
Path of Least Resistance
or Minimalism
or energy conservation,
so if there is a better and quicker way
of achieving results then we must use it.
Infact, Vedic Mathematics,
as just outlined here with the one-line method
of cubing numbers, utilising the formula:
$(a + b)^3 = a^3 + \underline{3}a^2b + \underline{3}ab^2 + b^3$
actually **GLORIFIES ALGEBRA**.
Mathematic Teachers need not be threatened
by this new, though ancient system, but rather
embrace and celebrate it,
and integrate it with the current system.

A POETIC SUMMARY OF "THE VEDIC MATHEMATIC SUTRAS AT WORK IN THE PHI-LOSO-PHI OF THE PHI PHENOMENON"

I AM
YOU ARE
THE LIVING EMBODIMENT OF PHI
in all its Beauty, Elegance and Grace;
I AM
WE ARE
its Breath and pulsing Base 12 encodement
I am the Privileged
MatheMagical Ambassador Of Phi
Numerical Nomad / Mathematical Monk
requesting invitation

to your Country, to your School,
and Central Nervous System
to share these Gems with you.
The Golden Mean is an application of
several Sutras, the foremost one being an ancient
Puzzle of <u>Proportion</u> called:

Anurupyena

like 1 is to 2 as 4 is to 8
(1 : 2 :: 4 : 8).
But Nature does not work
in perfect midpoint balancing
(symbolised by 1/2 or .5 or a half);
She idiosyncratically prefers a fraction more than
.5 called .618033 (1/φ)
a slight though very powerful sexual tension
that fusses on and cascades forever;
that way she uninhibitedly moves in whirligigs
generating tornadoes within larger tornadoes
as she pirouettes to a dodecahedral drum
of 12 paired beats.

She has evolved out of the Ultimate Proportion or
Golden Division of Unity
nothing that you can put your finger on,
no actual number that your lips could express
but more a principle of:

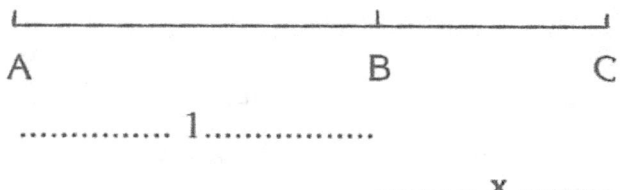

such that there is an equal and fair sharing
and proportioning of territory:
The Inter-Galactic Peace Treaty declares that:

$$\frac{AB}{AC} = \frac{BC}{AB} \quad \text{or}$$

$$\frac{1}{1+x} = \frac{x}{1}$$

The next Sutra says:
Vertically and Cross-Wise
a form of Cross-Multiplying:

this thing of the 'X-Factor'

this Right-Eye is connected to the Left Brain

and the Left-Eye is connected to the Right Brain

optical nerve cross-over enigma

Thus $\dfrac{1}{1+x} = \dfrac{x}{1}$

becomes

$x(1 + x) = 1 \times 1$

$x + x^2 = 1$

$x^2 + x - 1 = 0$

Another Vedic Formula or Sutra relating to
Differential Calculus
or
Calana Calanabhyam
determines

the dual roots of this Quadratic Expression,

surgically examining the coefficients,

and by mere mental dexterity and simplicity

evaluates: $x = \dfrac{-1 \pm \sqrt{5}}{2}$

$= 1/\phi \quad \text{and} \quad -\phi$

This is what makes wheels turn within wheels,

the Sphere to implode, the D.N.A to self-replicate,

the Phi Spiral to grow infinitely larger or smaller,

scale invariantly.

The internal Pulse of Phi,

the heartbeat pattern that sets the pace

for genetic recursion is based on the repetition of 24

Reduced or Digitally Compressed Fibonacci Digits

another Sutra called:
Digital Sums
or
Samuccaya
that pair themselves as 12 x 9

and have a sum of 108 !

OUR GENETIC 'SPHERICAL' MEMORY OF THE TETRAHEDRON AND THE CUBE.

(SECTION 1 OF 2)

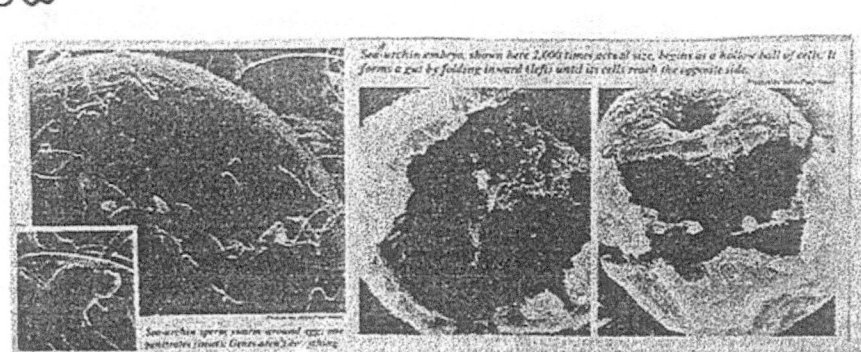

Derald Langham, an agricultural geneticist, showed with his microscope that a sea-urchin embryo starts as a single cell and is then dimpled by sperm to cause it turn into a torus, which is really the ability to turn itself inside out. This is shown in the top picture. Below, a mouse egg divides from a single cell or zygote to form 2 cells (The Monad becomes the Dyad), then 4 cells forming the Tetrahedron, then 8 cells to form the Cube.

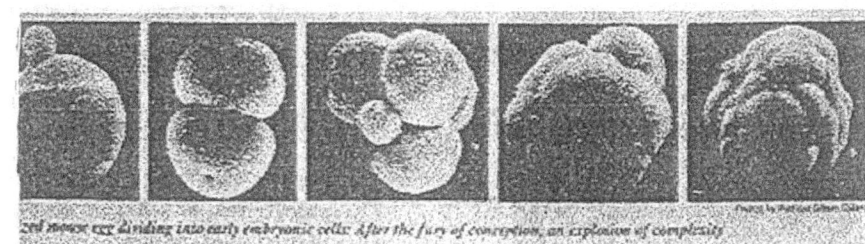

Phi, being a phenomenon of 12 and 24 will choose specific 3-D and 4-D and many multi-D models to transfer intelligently its living library of data:

As 12 Pairs of 9 it chooses

the <u>Dodecahedron</u>

of 12 pentagonal faces;

As 24 Reduced Fibonacci Digits

it forms or packs itself into

1) The <u>STAR-TETRAHEDRON</u> that has <u>24 Faces</u> and <u>24 edges</u>, and

2) The <u>CUBOCTAHEDRON</u> having <u>24 edges</u> and formed from the centres of <u>12 Spheres</u> surrounding a central 13th, the Christ Principle or

3) The <u>TRIAKIS TETRAHEDRON</u> composed of <u>12 triangles</u> which is the <u>Archimedean Dual</u> of the Truncated Tetrahedron

To be able to bridge the Dimensions, this set of infinitely recurring 24 digits must find a 4-D template to store its memory.

It chooses the 24 Squares of the HyperCube or Cube Within Cube a.k.a.

<u>Tesseract</u>.

Remember that these shapes are part of your <u>Genetic Memory</u> or <u>Creation Geometries.</u>

From the original ZYGOTE

(fusion of father/mother sex cells: Sperm/ovum)

the <u>ONE</u> Sphere became <u>TWO</u> Spheres became <u>FOUR</u> Spheres (the joined centers of 4 Spheres create the <u>TETRAHEDRON</u>) became <u>EIGHT</u> Spheres (the joined centres of 8 Spheres create the <u>CUBE</u>) became SIXTEEN Spheres or Cells that comprise the 16 Spheres of the 4th Dimensional Cube or <u>TESSERACT</u>.

Thus within the first Hours or Day of your holy Conception or Creation

you were all these familiar Geometric Shapes and linked to the many

<u>Angles of the Angels</u>

of the Many Dimensions.

But it doesn't end here, from 16 Cells you naturally

became THIRTY-TWO which became SIXTY-FOUR at which stage the geometry collapses and you became the

Tube Torus Doughnut

having 2 openings,
one becomes your Mouth,
the other your Anus. (Fig 35)

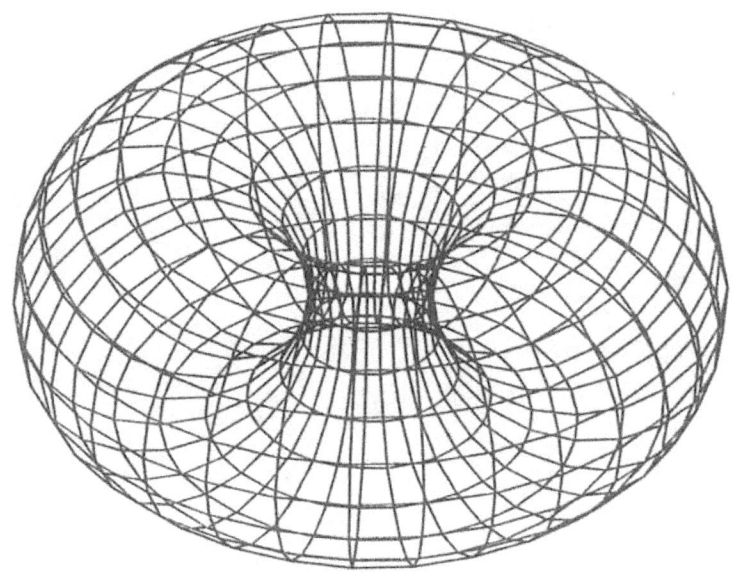

Fig 35
TUBE TORUS DOUGHNUT OF OUR CREATION GEOMETRY

OUR GENETIC 'SPHERICAL' MEMORY OF THE TETRAHEDRON AND THE THE CUBE.

MITOSIS: CELL DIVISION. THE 1 BECOMES THE 2.

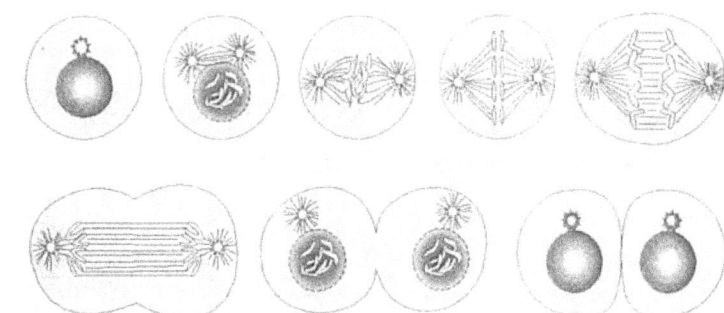

You may remember from your school days that organic chemistry is tetrahedronally co-ordinate, meaning that four spheres can be the arrangement of alike elements to stack or pack as a tetrahedron as shown below in Fig 36.

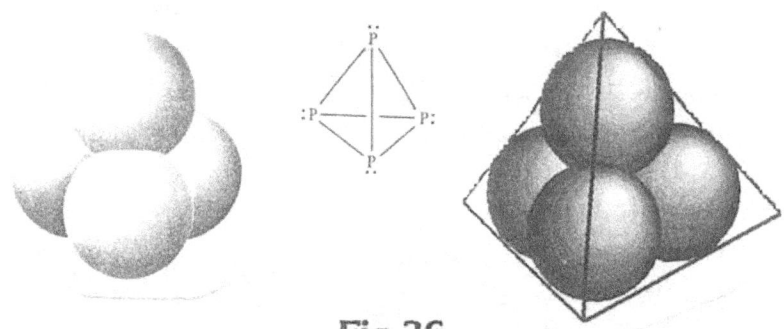

Fig 36
Four Spheres Touch to Form the Tetrahedron

Fig 36 also shows the structure of white Phosphorus molecules represented as P_4, a common allotropic form of the element.

You could also view the tetrahedron as emerging from the geometry of traced lines between the centres of 4 touching Spheres, Fig 37.

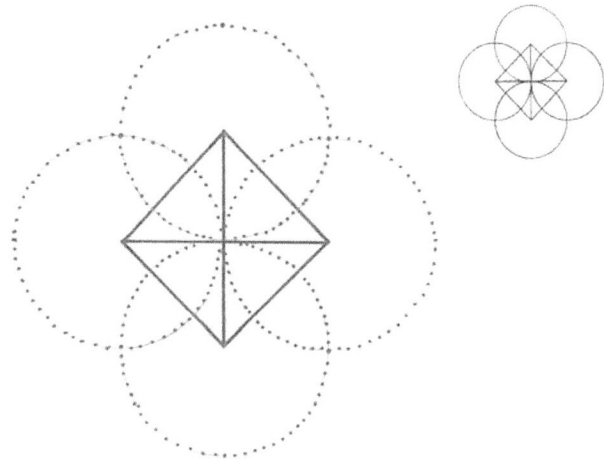

Fig 37
4 Spheres Touching. A Top Down View of Tetra.

The tetrahedron is the form containing the least volume for its designated surface area, accounting for its ability to withstand great external pressure. Often you will see tetrahedral milk containers for this very reason. (Whereas the Sphere has the reverse meaning in that it has max   volume for minimum surface area, as seen in soap bubbles and as storage tanks that tolerate maximum internal pressure).

Fig 38 shows a tetrahedral frame made of wire that when dipped in a soapy mix shows clearly the geometry of minimum forces, minimum volume and maximum strength. This is expressed in our first hours of life when we became the Tetrahedron as explained in the previous pages. Nature also expresses this in the creation of living microscopic organisms called Nesselarian, shown also in Fig 38, its skeletal view which is composed of silicon.

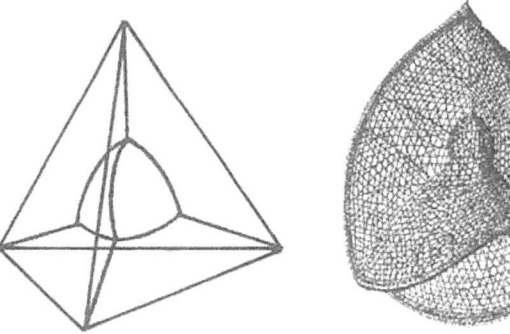

Fig 38
Soap Bubbles and Nesselarian

Why do you think Nesselarian has a silicon skeleton? The outside is the Inside. If you study the molecular structure of of SiO_2 (Silicon Dioxide found in glass and quartz), in Fig 39, you will see that every silicon atom is located at the center of a tetrahedron with the Oxygen atoms at its four corners.

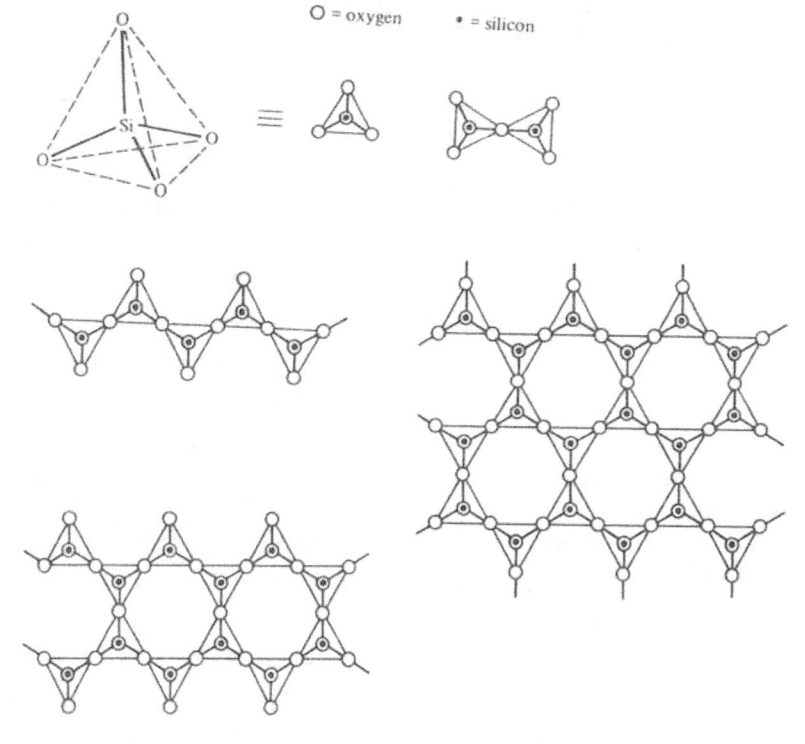

Fig 39
Tetrahedral Form of Silicon DiOxide (Quartz)

Shown also, in Fig 40a, are other tetrahedral molecular forms of methane CH_4, ethane C_2H_6 and Ammonium NH_3 the foundation of amino acids that function as the building blocks of life. In Fig 40b observe the 109.5° internal bond angle of the tetrahedral Alkanes which are Carbon based.

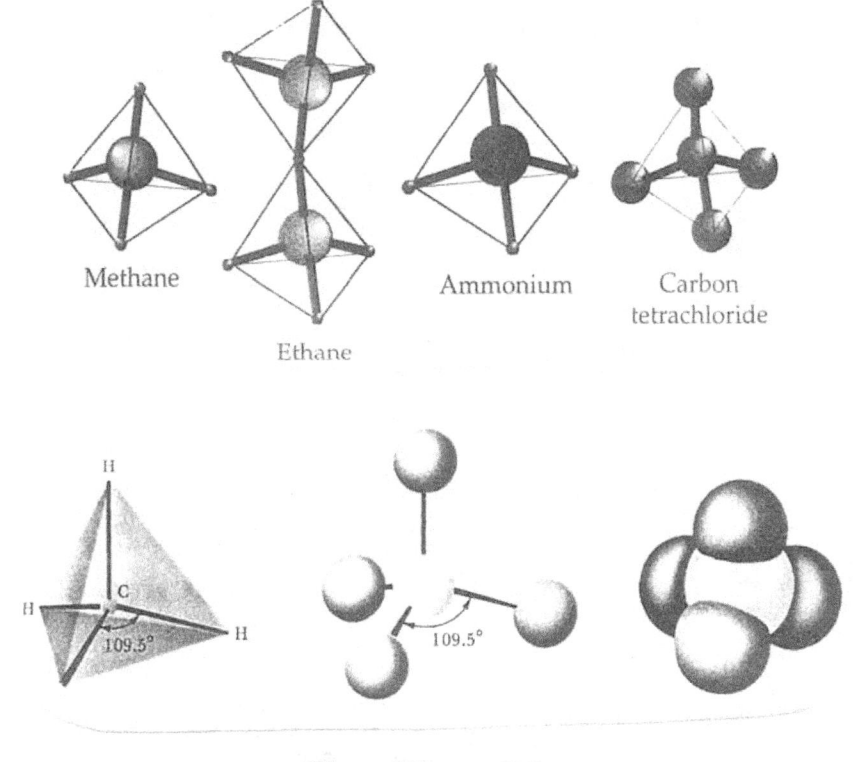

Figs 40a and b
3-Dimensional Tetrahedral Arrangements

Whereas Diamond is composed only of Carbon atoms in a tetrahedral net with short and strong carbon-carbon single bonds between the atoms. This makes it not only the strongest element but also the world's best conductor of heat, having a melting point of 3,500°, higher than that for any other element. Fig 41 shows the structure of diamond, a major allotropic form of carbon (Carbon has two major allotropic forms: graphite and diamond. Refer to Fig 27 for the hexagonal structure of Graphite).

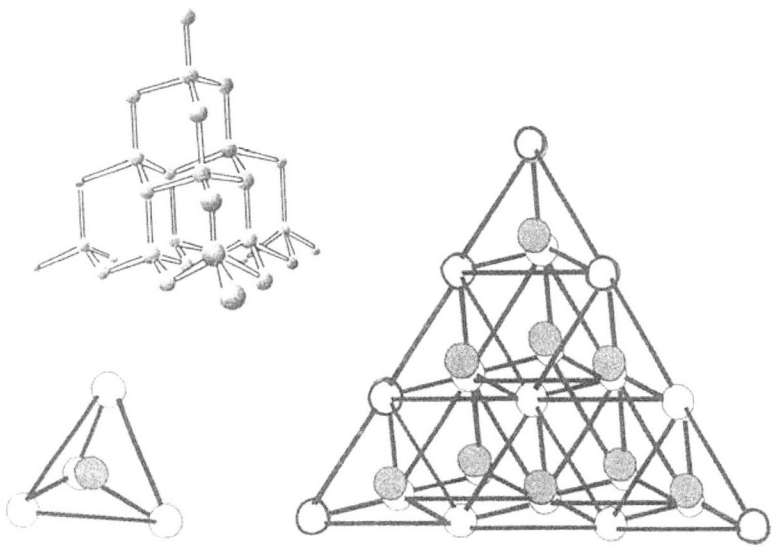

Fig 41
Tetrahedral Structure of Diamond's Carbon Atoms

With an understanding of your tetrahedral beginnings, here comes the important link: What is it that is <u>Tetrahedral</u> and also relates to the <u>Phi Code of 12 Pairings</u>? The answer lies in what nuclear physicists and mystics describe as THE MAGIC NUCLEUS. This is shown in Fig 42, taken from Professor Chris Illert's highly recommended book: "ALCHEMY TODAY" The diagram shows the tetrahedral geometry of the Helium nucleus which is also known as an Alpha-Particle (represented by the Greek letter α), but what is this Alpha Particle? and how does it relate to the 12 Phenomenon?

a nuclear FUSION reaction, as in our sun

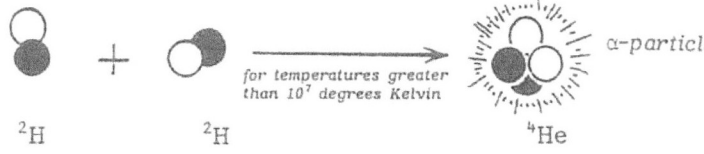

In this diagram we have drawn protons as black spheres and neutrons as white spheres, even though the nuclear binding force cannot tell the difference between electrically charged and uncharged particles.

Fig 42
The Tetrahedral Alpha Particle: Adyarium 12

The following is taken from Chris Illert's chapter: **'Platonic "Fire" And The Tetrahedron (the "s" shell)'**:

There are a number of super-stable atomic nuclei, including ^{4}Helium, (^{12}Carbon), ^{16}Oxygen, (^{40}Calcium), ^{56}Iron and ^{208}Lead, which we call **"magic"** (or "near-magic") **nuclei**. Other atomic nuclei are comparatively "unstable" and they undergo nuclear reactions ("transmutations"), either emitting or absorbing nucleons and other things, in order to become "magic".

The lightest of these super-stable ("magic") nuclei is ^{4}**Helium** (Fig 43) made from two protons and two neutrons arranged at the four vertices of a **tetrahedron. Hydrogen atoms** in our sun, being non-magic hence "unstable", burn (**fuse** together) to form Helium in the process of giving off huge quantities of energy which we recognise as **sunlight**.

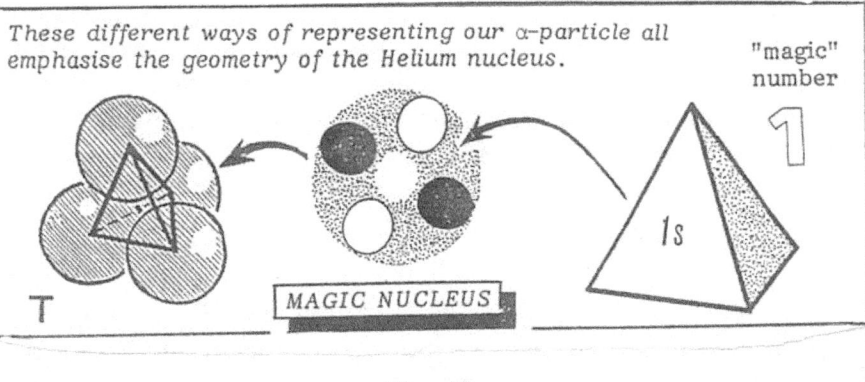

These different ways of representing our α-particle all emphasise the geometry of the Helium nucleus.

"magic" number

**Fig 43
Creation Of The Alpha Particle: ADYARIUM 12**

This is the basic nuclear reaction inside young suns made predominately from gravitationally collapsing clouds of Hydrogen. It was **Hans Albrecht Bethe**, in 1939, who solved this mystery of the sun's energy production, and he was awarded the 1967 Nobel Prize for this discovery.

The simplest nuclear reactions involve two heavy hydrogen nuclei that fuse together to produce a tetrahedral Helium nucleus (see Fig 43). Modern physicists call this "magic" ^{4}He (Helium) nucleus an **alpha-particle** (usually written as α-particle). Yet at the turn of the 20 century, before α-particles or atomic nuclei were known, **Occult Chemists** were systematically drawing and writing about a "chemical element" (which they called **Adyarium 12**) made from 4 spheres (nucleons?) arranged in a tetrahedron, each sphere being made from three smaller particles (quarks?)... **hence the number 4*3 = 12**. They also identified Adyarium 12 with one of the several kinds (isotopes?) of Helium, at least a decade before the existence of isotopes was suggested by **Soddy**.

As the four nucleons in an α-particle necessarily arrange themselves at the vertices of a tetrahedron, its clear that there must be a hole in the centre of the nucleus and for this reason we often represent the nucleus in two dimensions by a naive spherical-shell diagram. It is important to view this simplest "magic" nucleus as a shell rather than a solid sphere. On other occasions we might wish to emphasise the tetrahedral geometry of the nucleus and ignore the individual nucleons so we just draw a tetrahedron (above right in Fig 43). Because this first "magic" nucleus is made from only one tetrahedron (i.e. one α-particle) we say that it has a "magic number" of one Adyarium, (known as the **"1s" shell**).

Quantum mechanical descriptions place very little

importance on knowing the position and movement of individual nucleons within the nucleus and accordingly miss the underlying tetrahedral geometry discussed above. Instead, they use straight lines of "potential wells" that view the energy rather than the geometry of the atomic nuclei.

...Since initial "micro-psi" observations in the **1890's**, Occult Chemists had asserted that the tetrahedral "Adyarium" (otherwise known as an α-particle) was the fundamental building block for stable (nuclear) matter. For reasons of structural stability, and symmetry, based on sphere-packing considerations, **Buckminster Fuller** would have agreed as early as the 1930's...In fact, even as early as the 1890's, **Mme Curie** was finding that radioactive (large) atomic nuclei spit out entire α-particles, as if they existed as distinct entities inside nuclei before being ejected (thought at the time she didn't know that the radioactive emanations were, in fact, α-particles or **ADYARIA**).

After the first "1s" shell, Chris Illert's book goes on to describe a cubical star-tetrahedral shell of ^{8}Beryllium; the icosahedral shell known as the "1p" of ^{12}Carbon; and the dodecahedral nuclear shell as the "d" shell of ^{36}Argon.

With this understanding, you can now think of the atomic elements as the nesting of Platonic Solid Symmetries. Fig 44 shows that the micro-psi view of Oxygen as "T, I" which represents a Tetrahedron inside of an Icosahedron which was well described by Buckminster Fuller in his scheme of the two shells of Oxygen's nucleus.

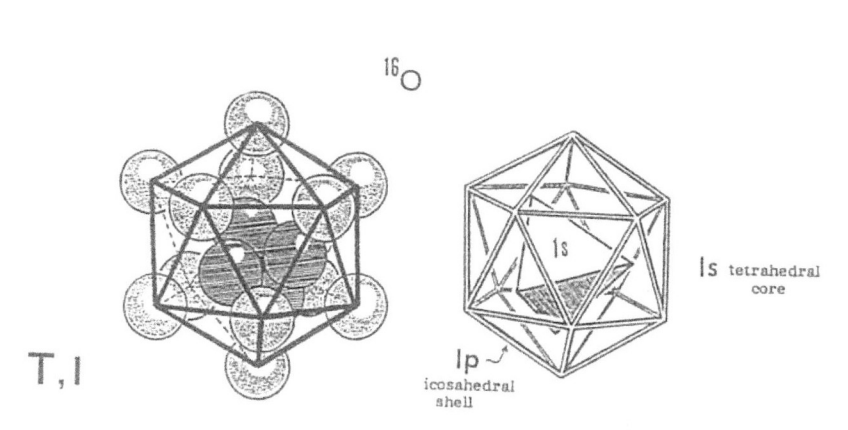

Fig 44
Oxygen's Atomic Nesting Of Tetra Within Icosa

Towards the end of this book I will link the two diverse worlds of Nuclear Physics and Magic Squares. The link or common language, as it will be seen, is something known as a <u>TETRAHEDRAL STACK NUMBER</u>, which is shown in Fig 45. This is really a solution in <u>IN THE NEXT DIMENSION</u>. It would be helpful for you to create lots of cardboard tetrahedrons and stack them as shown. When a row is completely stacked, it relates to the "<u>magic number</u>" of the "<u>magic nucleus</u>" and also with the "<u>Magic Square Constants</u>" or magic sums of the Magic Squares as revealed at the closing of this

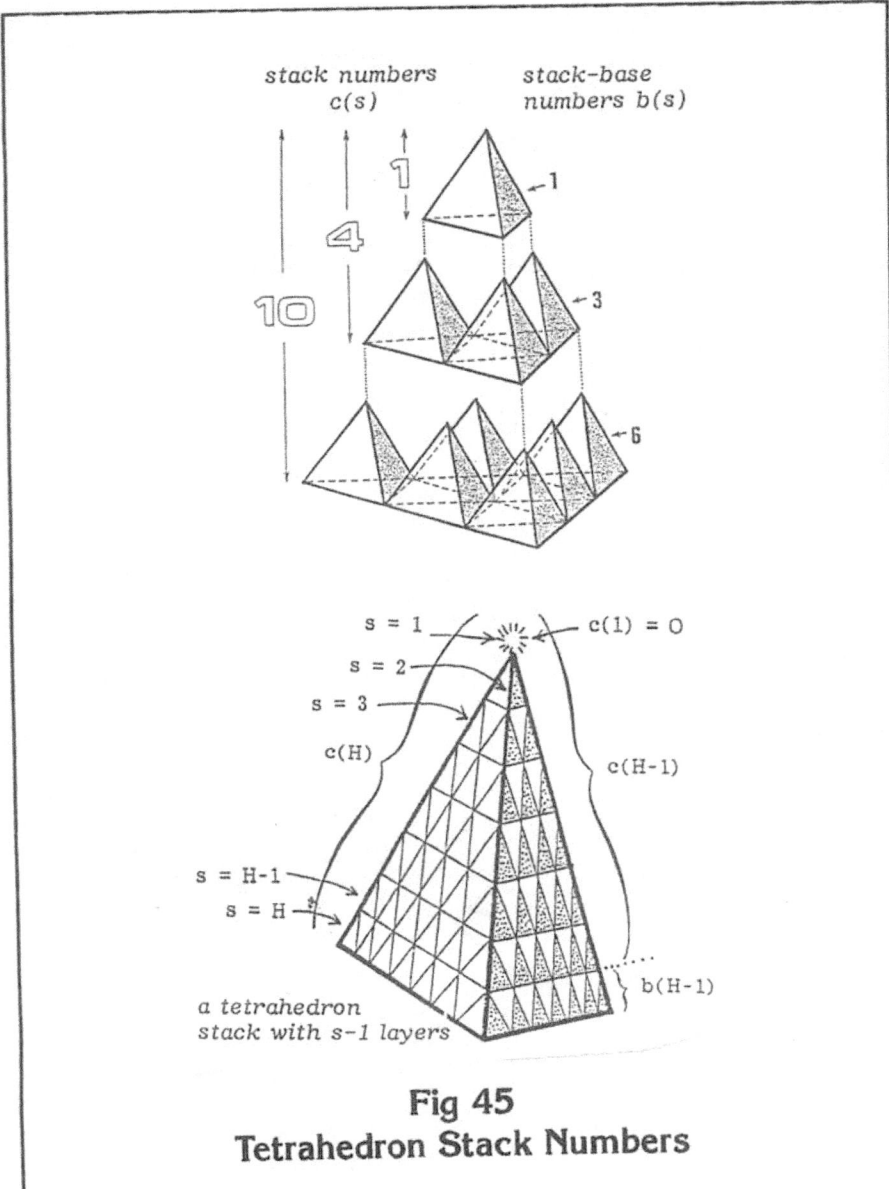

Fig 45
Tetrahedron Stack Numbers

book. I will also conclude this section on Platonic Solid Symmetries in the Periodic Table with another startling revelation linking the cubic atomic structure of Beryllium crystal to The Magic Square of 3 x 3 (See Fig 49a).

Of interest here is to present this material and arrive at the threshold of Alchemy (Fig 46). This can be realised by first looking at Buckminster Fuller's geometrical discovery of **THE JITTERBUG**, as shown in Figs 19 and 24. It has proven to the scientific world that atomic elements can elastically deform into the jitterbugging CubOctahedron (Fig 24). He showed that if you take the Icosahedron (having 20 Triangular Faces as shown in Fig 44) and half-twist the top and bottom in opposite directions, that it morphs into a shape with 12 Squares and Triangles (6 of each in this Archimedean Solid) known as the infamous CubOctahedron. It so happens that these Square facets can act as **"windows"** through which the aforementioned tetrahedral volume can escape. When ^{16}Oxygen nucleus gets excited it can "spit out" its core, as shown visually in Fig 46 and as written in the following nuclear reaction:

$$^{16}O \Rightarrow \Rightarrow \Rightarrow \Rightarrow {}^{12}C + {}^{4}He$$

What we have here is the creation or **Birth of an α-particle**. This is permissible as the "1s" tetrahedral

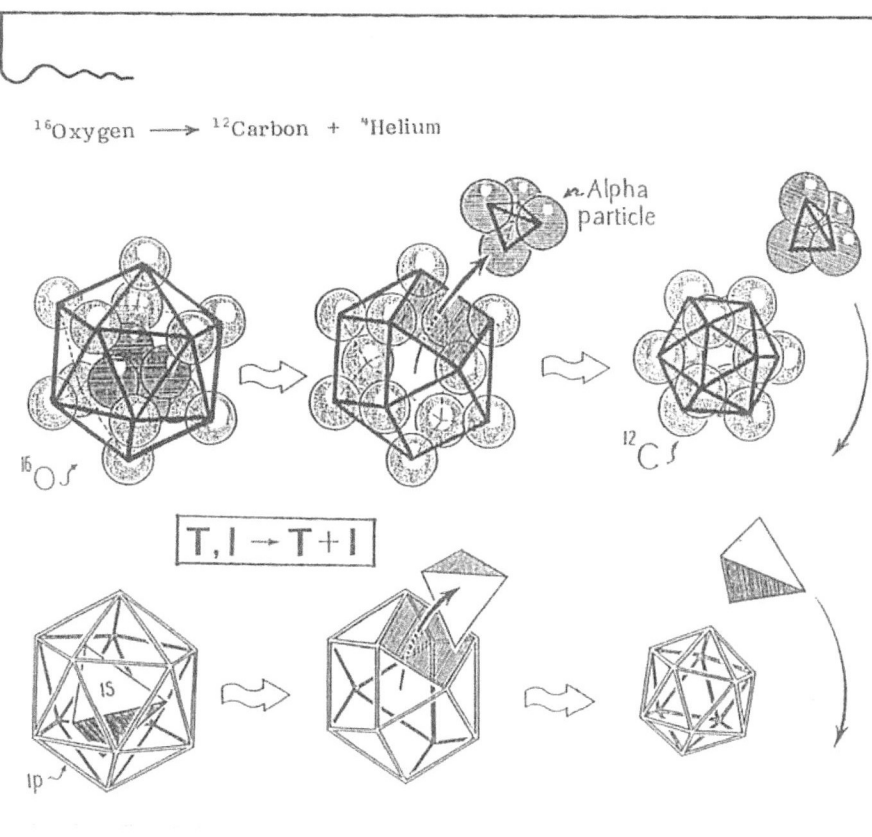

Fig 46
An Alpha Particle is Born. Jitterbug Transformation

core-shell is lighter than the surrounding icosahedral "1p" shell. Try enlarging and folding the following 3 nets or cut-out shapes to replicate this knowledge, as shown below in Fig 47.

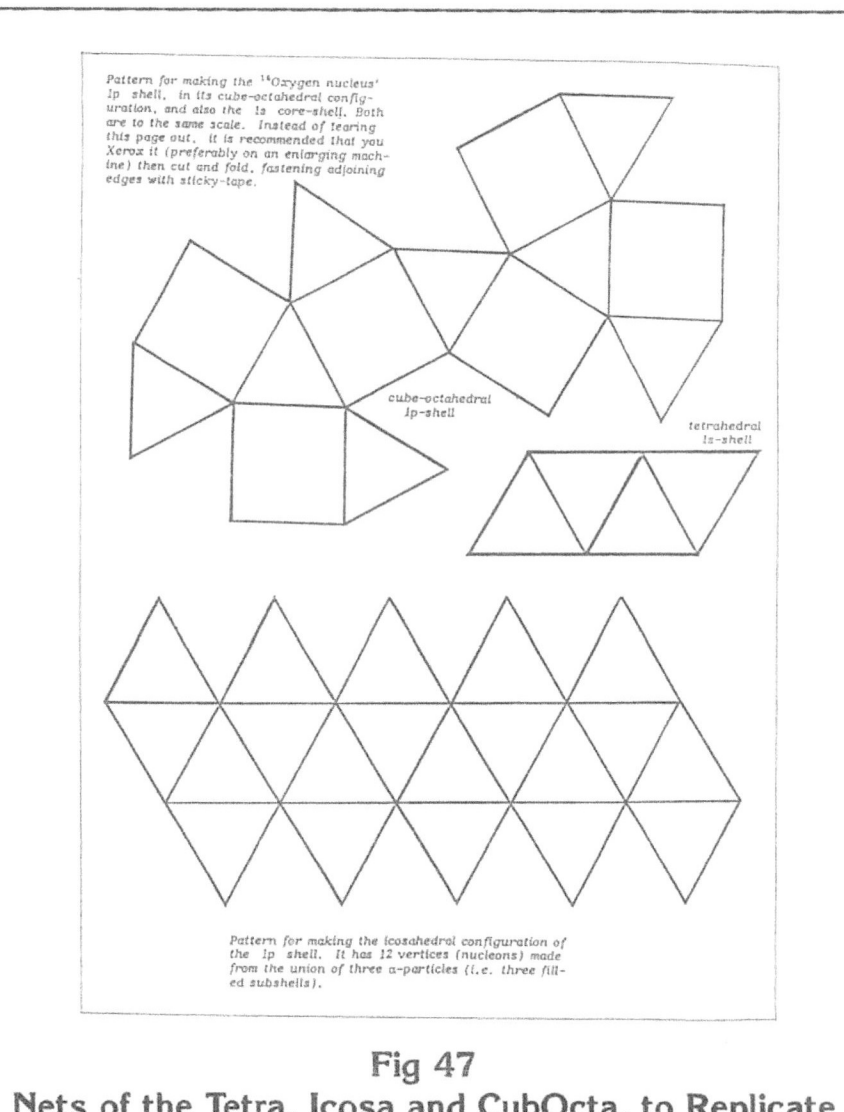

Fig 47
Nets of the Tetra, Icosa and CubOcta, to Replicate Fig 46. Copy and Enlarge these Nets on Cardboard.

If you accept that all the known and unknown Atomic Elements of the Periodic Table are but **nestings of the Platonic Solids**, and that their shapes can morph or jitterbug to create other entities, and that there is a grand or divine intelligence imbued in these Creation Geometries, then perhaps this will be a clue towards the understanding that Lead can be 'transmuted' into Gold. This will take you to the threshold of **ALCHEMY**'s door. ('Khem' was the ancient name for Egypt, from which we have derived the word 'Chemistry').

Before I go onto the CUBIC arrangement of our **8 ORIGINAL CELLS** of our Human Creation, I would like to present a fascinating discovery of mine that ties in the Occult Chemists view of the Platonic/Atomic matter and that of Magic Squares, and how this relates to: IN THE NEXT DIMENSION, a new and important mathematical "Vedic" Sutra. Have a look at Fig 48, again taken from Chris Illert's book. It basically says that Quantum Physics has blundered by describing Beryllium as a small Tetra inside of a bigger Tetra, which is nearly correct, but wrong. According to the Clairvoyant Scientists, Charles Leadbeater and Madame Besant, Beryllium is two equal-sized Tetrahedra that inter-penetrate or inter-digitate to form a Star-Tetra whose 8 vertices are really a Cube. What is of interest here is that the cube elongates, due to rotation about the centre of mass as shown by the arrow. Does this elongation approximate to the Golden Mean? and how else can we arrive at the same diagram? I believe that Magic Square Art is the "back door into Nuclear Physics". I can show how these naive 2-D magic square patterns are but shadows of the 3-D nuclear geometries. This has been explained in my book: THE BOOK OF MAGIC SQUARES Volume 3, exploring the relationship of a Magic Square of 3 x 3 (Fig 49a) to its Natural Square of 3 x 3 (Fig 49b).

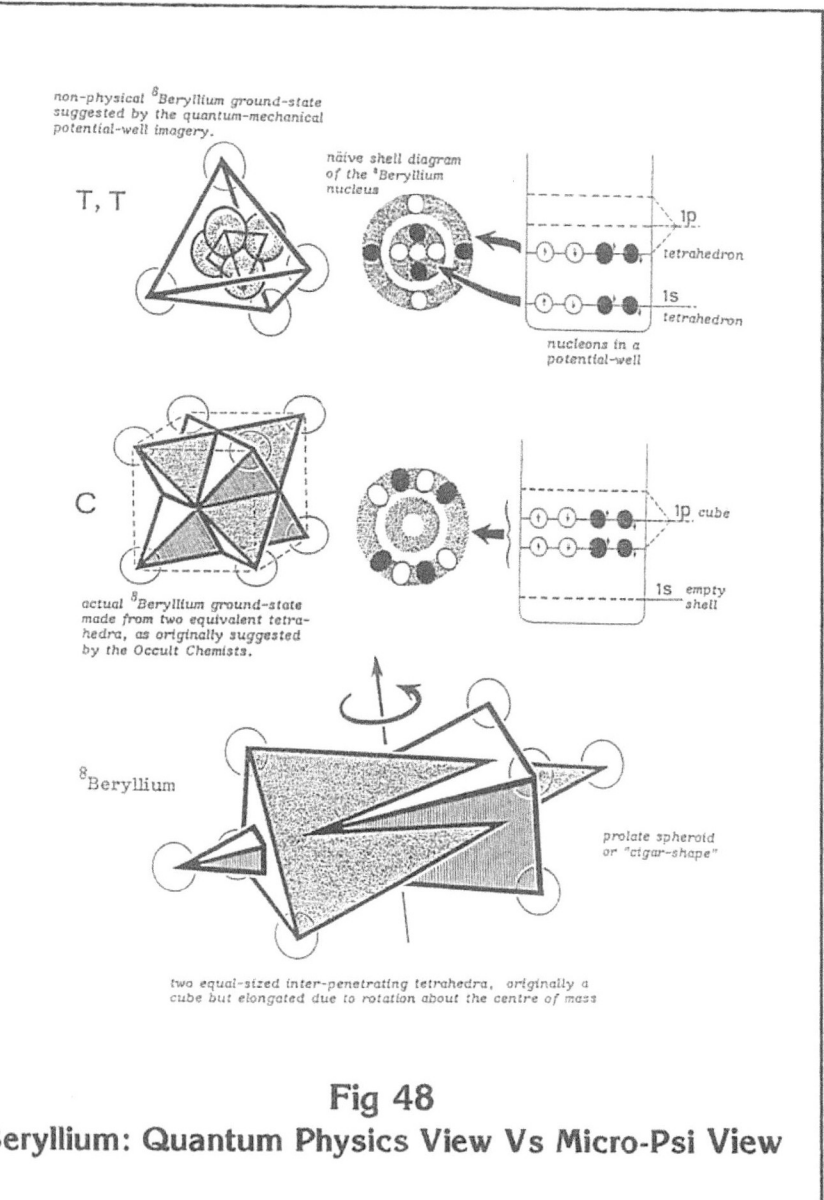

**Fig 48
Beryllium: Quantum Physics View Vs Micro-Psi View**

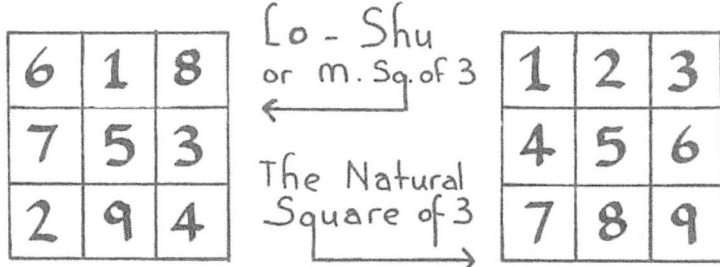

**Figs 49a and 49b
Magic Square and Natural Square of 3 x 3**

We are going to perform the task of "Row by Row Analysis" of each of the 3 horizontal lines or rows of the Magic Square of 3 x 3. Each row, Row 1: [6 1 8], Row 2: [7 5 3] and Row 3: [2 9 4] will be plotted in the Natural Square. First let us represent either the Magic Square (M.Sq.) or Natural Square (Nat.Sq.) as a configuration of 9 cells or 9 dots in a 3 x 3 matrix. Figs 50 a,b,c shows the result of asking:
a)- Where do the 3 numbers of Row 1: [6 1 8] appear or position themselves in the Natural Square of 3 x 3?
b)- Where do the 3 numbers of Row 2: [7 5 3]

appear or position themselves in the Natural Square of 3 x 3?

c)- Where do the 3 numbers of Row 3: [2 9 4] appear or position themselves in the Natural Square of 3 x 3?

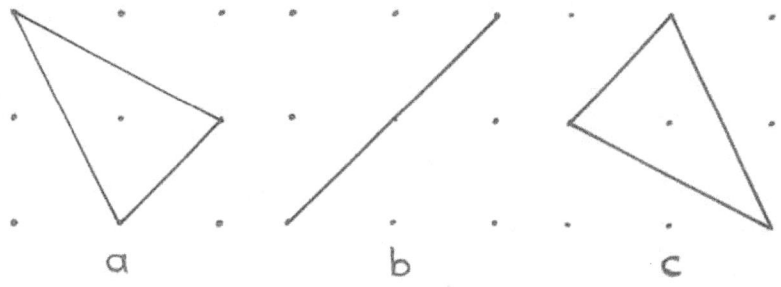

**Figs 50 a, b, c
The Row-By-Row-Analysis of the M.Sq. of 3 Substituted Into the Nat.Sq. of 3.**

So far we have 2 elongated triangles and a slanted diagonal line. What happens if we superimpose these three images upon themselves, as if they were drawn upon clear plastic transparencies.

Fig 51 shows the result of an elongated Star of Solomon that approximates the Golden Phi Ratio.

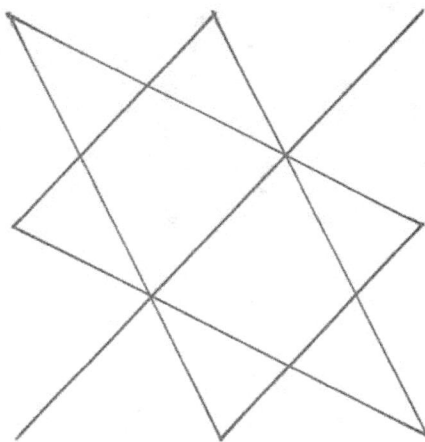

**Fig 51
Star of Solomon Formed from M.Sq. of 3 x 3**

Now if you were to look at Fig 51 with 3-D eyes you would perceive the two elongated phi triangles as two interpenetrating phi Tetrahedrons or phi Star-Tetra as shown in Fig 48, but also look at the slanted line in Fig 51 and correlate it exactly to Fig 48's spin of the cubic crystal Beryllium's elongation due to "<u>rotation about the centre of mass</u>". The similarity here between Magic Square Art and Nuclear Geometry can be no mere chance. If you drew Fig 51 again but superimposed the pattern upon itself at an angle of 90° you would be close to viewing King

Solomon's psychic Seal as shown in Fig 52a. Technically, in Magic Square Jargon, it is known as The Rabbi's Golden Star Yantram at 0° + 90°. ('Yantra' is the plural of 'Yantram', a Sanskrit word for "power diagram acting as 'transistor circuits' or channels conducting psychic essences"). Fig 52b shows "The Pentacle of Rabbi Solomon The King taken from an ancient bible of magical talismans.

Figs 52 a and b
M.Sq. 3x3 at 0° + 90°
Pentacle Of King Solomon

Fig 53: The ASHMOLEAN STONES

THE ASHMOLEAN STONES

Considered to be pre-Plato or some texts say 10,000 years old, these stones, currently in the Museum in Oxford, are hand-sized carvings of the 5 Platonic Solids and some Archimedean Solids, like the CubOctahedron. In Plato's "Timaeus": a cosmic dialogue based on these Platonic Solids, Tetra is Fire, Cube is Earth, Octa is Air, Icosa is Water, but of high importance is the FIFTH ELEMENT (Dodeca), its 12 faces being representative of Prana or the Universal Life Force or the **Phi Code of 12 Pairs**. Keith Critchlow (Magic Squareologist and Phi Architect and Lecturer) concludes that:

"The essential forms and numbers then act as the interface between the higher and lower realms".

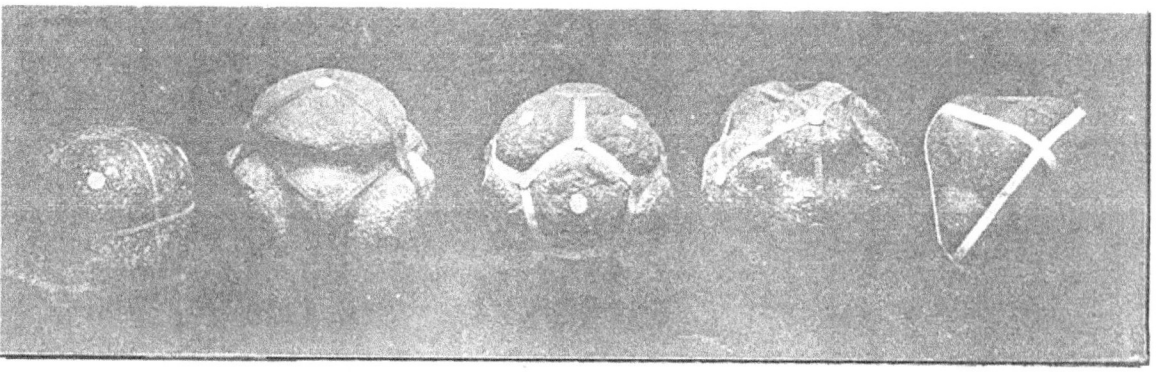

MORE NOTES ON PHI

If the greater part of a Line is of the length Φ and the lesser part is 1, then:

$$\frac{\Phi + 1}{\Phi} = \frac{\Phi}{1}$$

ie: $\Phi^2 = \Phi + 1$ or $\frac{1}{\Phi} = \Phi - 1$

In other words, it is **SQUARED by adding ONE**
$(1.618)^2 = 2.618$
(which is shown as Points S and T in the upcoming Fig 56);

and its **RECIPROCAL is found by subtracting ONE**:
$$\frac{1}{\Phi} = \frac{\sqrt{5} - 1}{2}$$
$(1.618 - 1) = .618$
(Point K in Fig 56)

THE POWERS OF PHI
(SOMETIMES CALLED THE LUCAS NUMBERS) IN TERMS OF THE FIBONACCI NUMBERS

In my previous book, THE BOOK OF PHI Volume 1, I showed the FIBONACCI SEQUENCE:
0, 1, 1, 2, 3, 5, 8, 13, 21, 34, 55, 89, 144, 233, etc.

Observe these numbers appearing in the Powers of Phi:

$\Phi^2 = \Phi + 1$ $= 2.618033$

$\Phi^3 = 2\Phi + 1$ $= 4.236060$

$\Phi^4 = 3\Phi + 2$ $= 6.854085$

$\Phi^5 = 5\Phi + 3$ $= 11.09013$

$\Phi^6 = 8\Phi + 5$ $= 17.94420$

These Lucas and Pyramid Numbers are highly respected in the world of Wall Street, according to Bryce Gilmore in his "Geometry of Markets".

This observable pattern demonstrates that each Power of Phi is the sum of the 2 previous powers: observing that the coefficients of Phi forms the Fibonacci Sequence over again, as do the whole number or integer parts of the Powers.

INTRODUCING THE "x" FACTOR

It is considered important to understand that Phi is a relationship superior to midpoints. Midpoints are a supposed equilibrium between two ends AB as shown by the point P in Fig 54 below. Actually, a midpoint between two tensions AB creates a standing wave or a stasis of inertia. That means a state of non-movement. This is good if that is what you require, but we desire to travel multi-dimensionally, and scale invariantly and the ideal relationship is not a halving or midpointing or equistasis as most people imagine, but a slight movement off-centre, a fraction more than half or .5 which magically generates an intense sexual implosion that allows movement to occur, that allows infinity to be met, that allows the mathematical equations of the atom to be the same as those of the galaxy. Fig 54 shows clearly that point P is not a desired division, whereas the points L and K are the desired outcomes represented by .382 (the reciprocal of phi squared) and .618 (the reciprocal of phi) respectively. (see Fig 56).

This diagram therefore assists you in understanding the mystical division of Unity where all waves are in <u>concrescence</u> (from the the Latin word: "Concrescentia" = a condensing, a compression, a coalescence or growing together).

If we were asked to measure this golden division of AK to KB, if our line AB = 1 we would have no obvious tools as to how to achieve this mystical measurement, but fortunately an Arab by the name of **Al Jabr** (Al Khwarizmi circa 825AD) introduced the modern concept of "Algebra": it being named after him. His simple, though history-changing introduction of: <u>"x" as the unknown factor</u>, coupled with the concept of <u>Negative Numbers</u> allowed the modern world to create technology and fly rockets to the moon. This is how this book started; we sought to analyse the internal division of unity (at point K) and

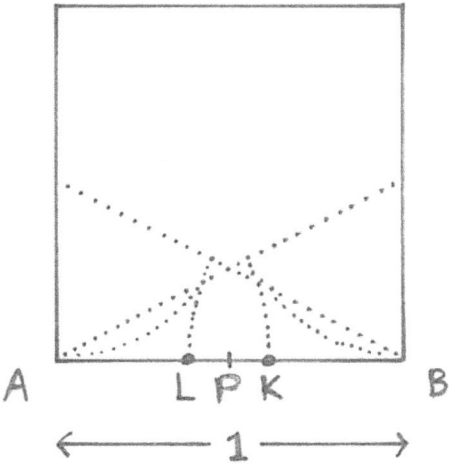

Fig 54
Reciprocal of Phi as a Fraction More than Midpoint

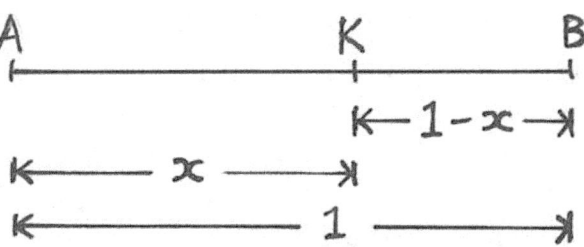

**Fig 55
Introducing the "x" Factor**

came up with an algebraic expression of:
$x^2 + x - 1 = 0$ giving the answer of .618.

It is a simple though brilliant application of human intelligence that if AB = 1 and AK = "x" the unknown factor, then the length KB must be Unity or 1 less this "x" factor and is expressed as '1 – x'. We started off our journey not knowing what value "x" is and conclude by knowing:

that $\dfrac{KB}{AK} : \dfrac{AK}{AB}$ which introduces the "x" factor:

$$\dfrac{1-x}{x} = \dfrac{x}{1}$$

The cross-multiplying of terms gives us the quadratic expression $x^2 + x - 1 = 0$. (For a detailed account of how how the golden mean was achieved, read the beginning of this book).

The point K, the Golden Mean Division of Unity AB is really an expression of the cosmic Trinitarian principle that knows how to harmonise extreme tensions or polarities. It is the <u>TriVision of Unity</u>.

If AB are polar opposites or extremes, like: Hot /Cold, Day/Night, Male/Female, Past Future etc then the Golden Mean at point K is the warm, dusk-like, androgynous zone where magic or Love happens in the Now!

On the following page, shown in Fig 56, is a combination of the internal and external divisions of Unity's Square AB, using circles to link the aesthetically pleasing distances between the 4 Roots "H and Q" (External Division: $x^2 - x - 1 = 0$) and "K and M" (Internal Division: $x^2 + x - 1 = 0$).

Such diagrams are the ground plans for temples from the past and inventions for the future.

You may know of the Vesica Piscis as the coalescence of 2 equally-sized circles in such a way that the circumference of one circle touches the heart or centre of the other circle. This was shown in Fig 1a, the first diagram that begat this book! illustrating the progression from the cube to the double cube to the golden section or division.

What we now have in Fig 56 is a very interesting variation of the Vesica Piscis, based on the Golden Section. It shows the meeting of 2 circles as they greet or kiss each other at the auspicious distance of Phi. It is for this reason that it is called the **PHI VESICA**. The 2 intersecting circles can be seen as ML and HR.

Another realisation is that the diagram of Fig 56 is simply the cross-sectional view of the tube-torus-donut (shown in Fig 14a and Fig 35), a 4th Dimensional Sphere technically known as The HyperSphere. Many scientists are currently excited about this model or toy as it explains magnetic fields and births of galaxies etc. Remember one of your childhood toys known as The Slinky, the D.N.A.-like spiral coil that could walk down steps. If you held its both ends and joined them into a loop or circle you would have created a Torus (Tori = plural). Like the HyperCube which is a Cube within a Cube, the Torus or HyperSphere is a smaller Sphere within a bigger Sphere, having an extra axis or 4th axis of spin. Slicing this donut in half would reveal the INFINITY Sign as an idealised Figure 8 lying on its side or depicted as 2 equal circles touching, side by side:

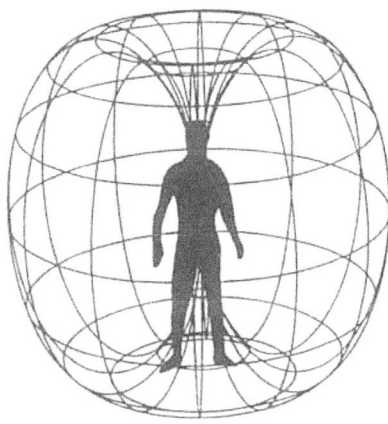

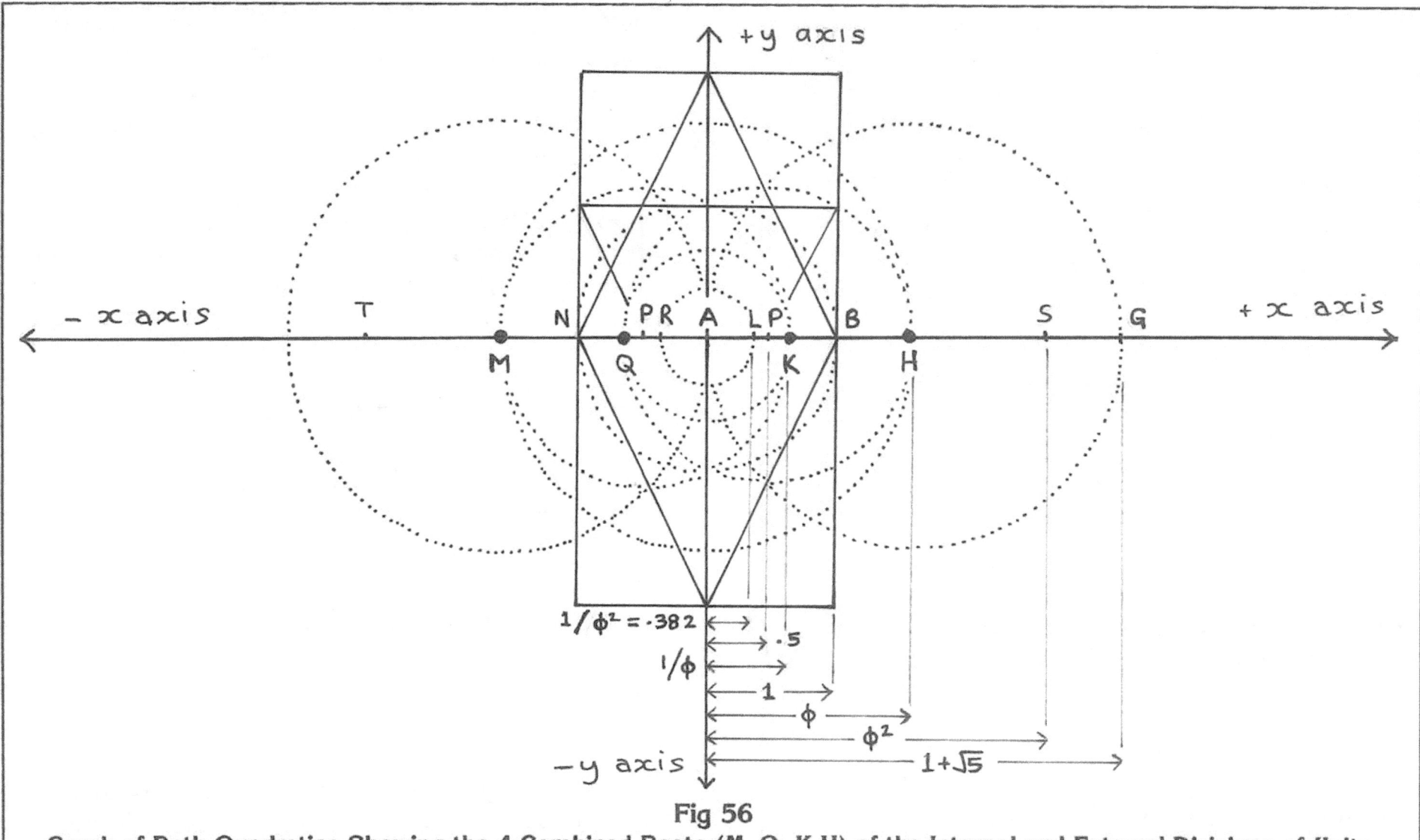

Fig 56
Graph of Both Quadratics Showing the 4 Combined Roots (M, Q, K, H) of the Internal and External Divisions of Unity.
GENERATION OF THE PHI VESICA. nb: The Cross-Sectional View of the Torus (Figure of Eight on its side)

The diagram in Fig 57 shows our desired geometry of nesting Tori, the ultimate fractal that is the emblem of the Infinite or **IN–PHI–NET**.

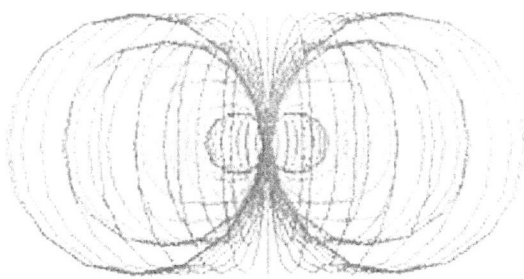

**FIG 57
NESTING TORI**

This dissertation is to put forward the same concept of nesting Tori but in a Phi relationship as shown clearly in Fig 57 and evolved in Fig 56. I believe that nesting tori in phi vesicae or 1.618 orbitals or shells is the true merkabah or spiritual light vehicle. Even to just imagine the Sphere or soap bubble as a sophisticated interdimensional space/time vehicle is correct thinking. How is it that Spheres and HyperSpheres can relate to higher dimensions.? To answer this, we would need to examine the Volume of a regular Sphere (which is $4/3\pi r^3$) and compare it to the Volume of the HyperSphere ($2\pi^2 r^3$). Did you notice that there is an extra "pi" in the formula of the HyperSphere? which accounts for rotation in an additional dimension. That it why it relates to: IN THE NEXT DIMENSION.

It would help you to understand this by ordering from America one of these HyperSpheres by getting onto the website: http://www.hypersphere.com/hs/info.html and then playing with it, observing how the geometry can turn itself <u>inside-out</u> which is one of the features of the 4th Dimension!

magic squares and hypercubes

Before I begin part 2 of "Our Genetic Spherical Memory of the Tetrahedron and the Cube" (see Figs 35 and 36) which concerns more the geometry of the 8 Original Cells or CUBE, I will introduce the notion of how 2-D Magic Squares can be converted into Tori. Magic Squares in the next dimension! It is of relevance here as I will show how Magic Squares can collapse into HyperCubes that have 24 Faces and this relates to the 24 recurring digits of the Phi Code as shown at length in THE BOOK OF PHI Volume 1, 2002.

(All Magic Squares also have corresponding Magic Cubes, in the next Dimension, and keep nesting in 4-D, 5-D, 6-D etc. I will show, towards the end of this book, how 20 years of research of the Magic Sums of the Magic Squares led to the discovery that all this data, if it had to be compressed into a suitable 3-D form or vessel, would collapse into THE TETRAHEDRON, or more specifically as Triplic Layers of the Tetrahedral Sphere-Packing Numbers!).

Fig 58 shows 1 of the 880 possible Magic Squares of 4 x 4, which unfortunately has the strange name of being called a "Diabolic Magic Square of 4 x 4", because its broken diagonals also add up to the sum of 34:   eg: $5 + 8 + 12 + 9 = 34$.

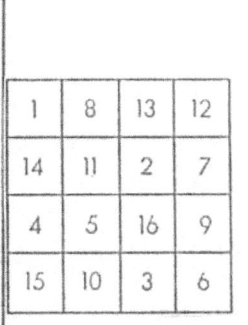
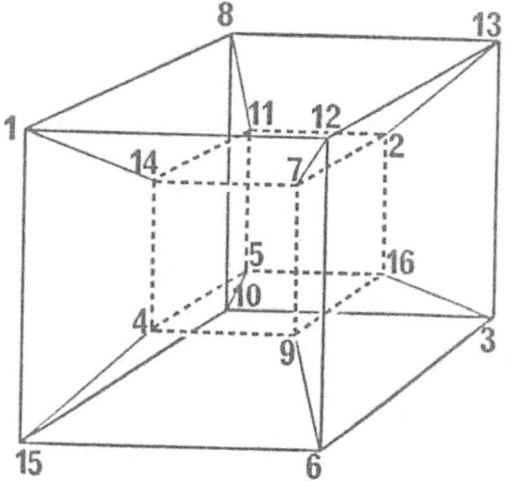

Fig 58
Diabolic HyperCube
and One of its 384 Diabolic Magic Squares of 4x4

It also means that it has the special property of being able to move or interchange any column or row from side to side without destroying its integrity or magic sum of 34 which is the Gematria or the mystical numbered-word for the holy name of the Lord "Adonai" which adds up to 34 in the ancient Hebrew! Magic Square Sums are really the secret-code names for the many names of Allah.

Study Fig 58 and observe how the 16 numbers or cells of the Magic Square can be morphed into the HyperCube! by transferring the 16 cells of the Magic Square to become the 16 corners of the HyperCube. (see Fig 12 for good graphics of this 4-D Cube or Tesseract). The result is incredible, for now the sum of the four corners of each of the 24 square faces of this HyperCube have the equal sum of 34. It had been shown that there are 384 different HyperCubes that can be formed, which includes rotations and reflections.

(These hypercubic properties were first illustrated by mathematicians J. Barkley Rosser and Robert J. Walker, both of Cornell University, in a paper published in 1938. Fig 59 is also by them. Both diagrams were subsequently reprinted in Martin Gardner's "More Mathematical Puzzles").

I suggest to the physicist in search of cosmic antennae to follow the path of this HyperCube with a long thread of copper wire, beginning at Number 1 → 2 → 3 ... → 16 and integrate this 3-D / 4-D sculpture into your circuit board.

Fig 59 shows another Diabolic Magic Square of 4 x 4 which other books may classify as "Pandiagonal" or "Nasik" (a place in India where the carvings of such were first observed by Westerners). It is the earliest recorded 4th Order Magic Square, found in an C11th or C12th inscription at Khajuraho, India, world famous for its prolific tantric stone carvings.

Mathematicians Rosser and Walker exhibited new properties of this square hitherto unknown. In a dramatic diagram (Fig 59) they showed how by simply bringing together the top and bottom of the square to make a cylinder then stretch and bend the cylinder into a Torus, they could make all rows and

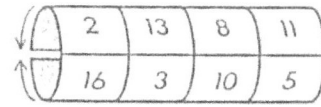

Fig 59
M.Sq. of 4 Morphed Into The Diabolic Donut

columns and diagonals to be now described as closed loops on the surface of the Torus Donut. Try starting at any particular number or cell and move along two squares away in any direction along a diagonal and observe how you will always arrive at the same cell you started with! This means that every loop of 4 cells, diagonally or orthogonally (ie. at right-angles or perpendicularly) will sum to 34. Also any square group of four cells will sum 34.
(Dictionary meaning of Torus or Toroid = In Geometry, a surface generated by the revolution of a conic (esp a circle) about an axis lying in its plane).

I hope this has been of interest to you.

"<u>ONLY HE WHO IS FAMILIAR WITH GEOMETRY SHALL BE ADMITTED HERE</u>"
(An inscription written over the entrance to the Platonic Academy).

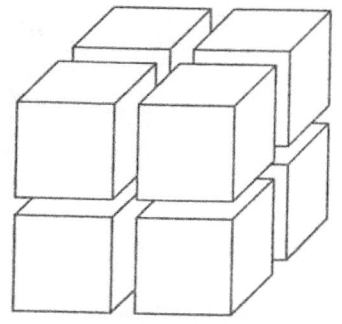

"<u>Wherever there is Number there is Beauty</u>"
Proclus
(410 – 485AD).

PART 2: SECTION 2

OUR GENETIC 'SPHERICAL' MEMORY OF THE TETRAHEDRON AND THE CUBE

Fig 36 and following portrayed the relationship of our first 4 cells of life to the divine physics of the Tetrahedron as the template or blueprint of all creation. This was Part 1. Part 2 of this chapter will focus on The CUBE and our identity of the 8 Original Cells of Creation which is cubic or really two interpenetrating Tetrahedra.

How is it that the Phi Code (12 Pairs of 9) relates to this chapter regarding The Cube? The connection is obvious when we ask ourselves: What 3-Dimensional Form has the quality of 12ness? The answer is: The Cube has 12 Edges and so does its Dual: The Octahedron has 12 edges, shown in Fig 60 and previously in Fig 14b. (See the Appendix for

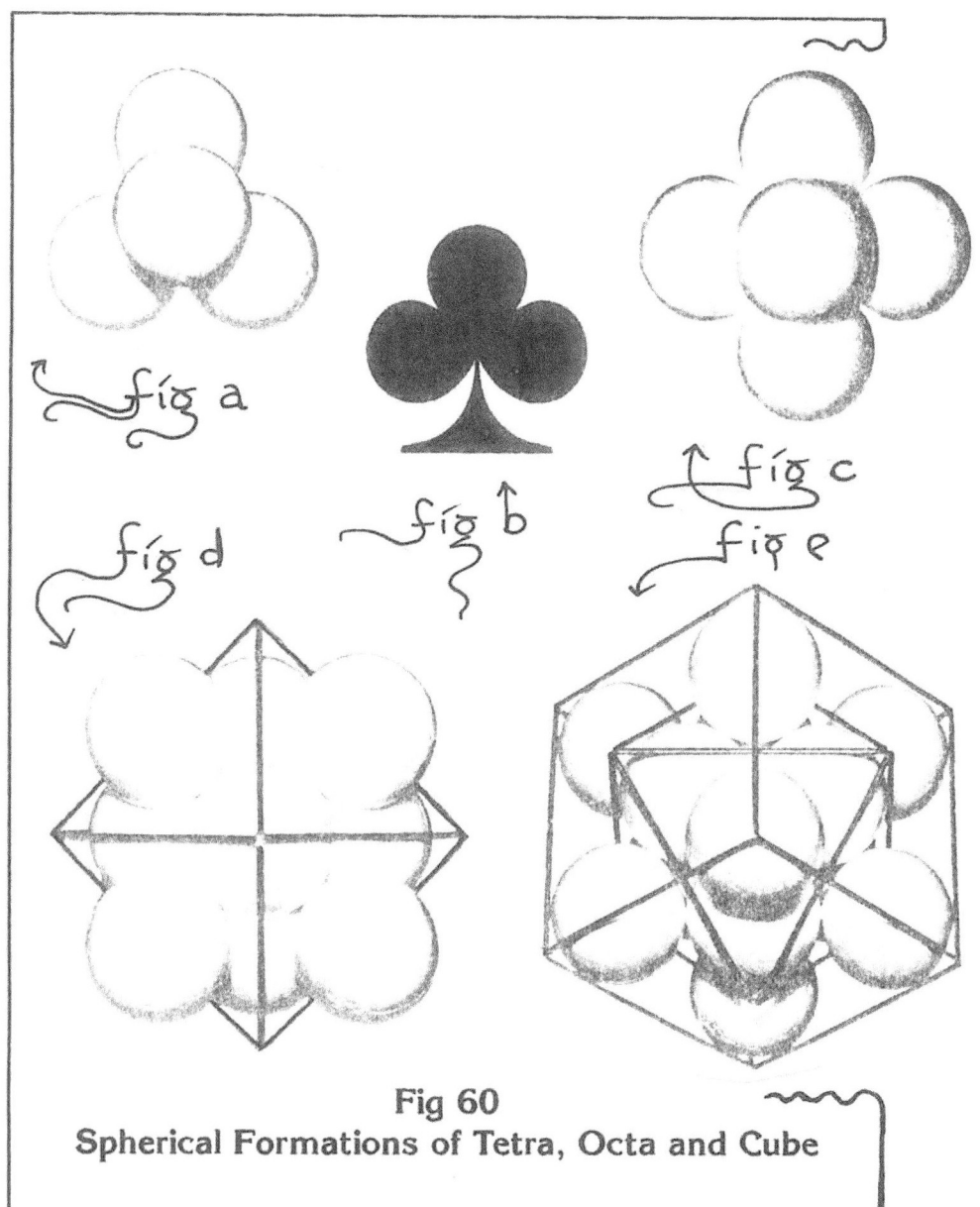

Fig 60
Spherical Formations of Tetra, Octa and Cube

graphics of the Duals of the 5 Platonic Solids).

Fig 60a shows the Tetrahedron as 4 spheres, but it is shown here to illuminate its outline of the Trefoil ("TRIFOLIUM": Latin for 3-Leafed) which is commonly seen in architecture but best known as the <u>Suit of Clubs</u> in the traditional playing cards (Fig 60b). Thus, the 2-Dimensional Trefoil or sign of clubs is really, <u>In The Next Dimension,</u> the famed <u>Tetrahedron</u>. Or conversely, the <u>Shadow</u> of the Tetrahedron is the Trefoil. I think it is important to think like this as it will strengthen your mental ability to think and feel in the higher dimensions.

Fig 60c shows 6 spheres forming the Octahedron (the 3-D Form of the traditional playing cards' suit of Diamonds!).

If 8 spheres are placed upon the 8 triangular faces of the Octahedron, the efficiently cubic-packed form of Fig 60d is created and is reminicient of how the Silicon-chip crystal molecule is cubically or star-tetrahedrally arrayed which is shown in Fig 61. Notice also in Fig 60e how the Cube emerges around these 8 Spheres.

Why do you think the 8 Spheres that form the Cube or the 8 vertices of the Cube have been traditionally likened to Jesus' Crown of Thorns that hangs on The Cross? Again, we have to think In The Next Dimension by asking ourselves another

Fig 61
Atomic Structure of Silicon Chip Crystal:
Cubic or Star-Tetrahedral

question:

What if we were to fold down the 'walls' of the sacred Cube, to look inside its 'skeleton', what shape would we see? Fig 62 shows clearly that the Christian or Gnostic Cross is the solution to another Shadow of another Multi-Dimensional Form. Lawrence Blair, in his book on sacred geometry "Rhythms Of Vision" explains how many of the early Gothic cathedrals, architectured by the Masons reflect the structure of this cubic figure. The Masons hid and incorporated their great knowledge of sacred proportions in their majestic buildings.

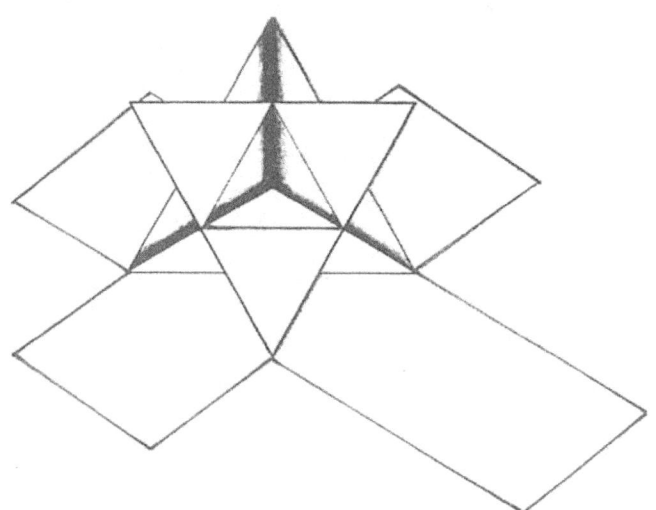

Fig 62
Walls of Cube Laid Down Exposing Mystic Cross

Is there a simple and convenient 2-D picture that can express the Cube within a larger Cube?

Fig 63a shows another conceptual view of the Doubling of the Cube. Remember how we began this book, in an attempt to extract the roots of the Golden Mean by beginning with the Square and Doubling it to get 1.618033. Well here is a beautiful fractal image of the Micro within the Macro, where 6 Circles are tangent to a central Seventh. <u>Notice that the highlighted smaller Cube in the center is the **Square Root** of the larger Cube</u>. Fig 63b shows again the idea that a 2-D form, like the Hexagram of 2 interlacing Triangles, can take us into the next dimension by manifesting cubic space. Thinking like this, what do you think the 3-D expression is for the 6 circles tangential to the central seventh is, as depicted in Fig 63a. The answer is shown in Figs 24a and b: the twelve spheres around the central thirteenth, the 12 spheres relating to the 12 pairs of 9 in the Phi Code.

The Sphere is the ultimate Space Craft or Merkabha Vehicle or Inter-Dimensional Form. I suspect it is because mathematically, a Sphere which is doubled in its radius is squared in its plane area in section and cubed in its volume!

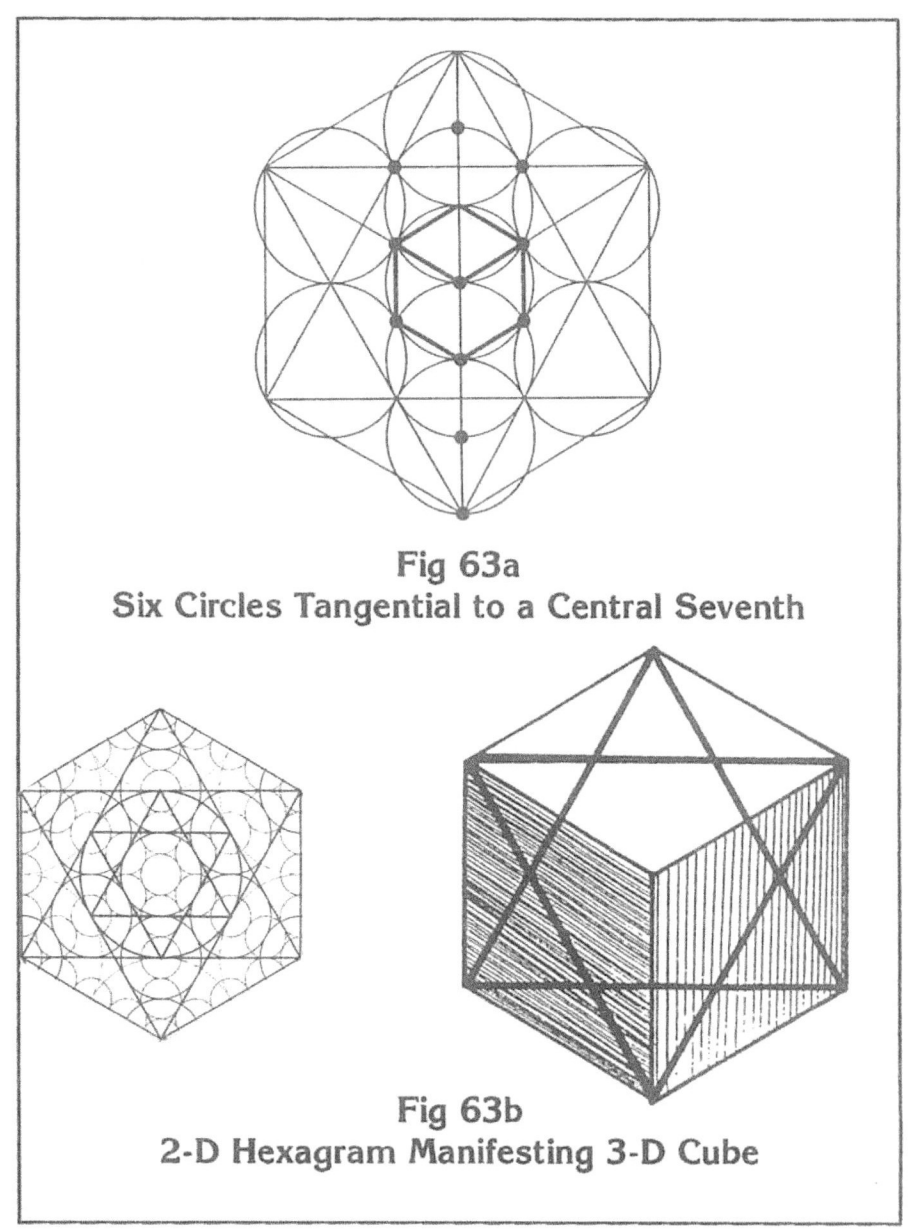

Fig 63a
Six Circles Tangential to a Central Seventh

Fig 63b
2-D Hexagram Manifesting 3-D Cube

This notion, describing the evolution of a straight line, to a plane, to a form (radius, area of circle, volume of sphere) is the laddering of dimensions defined really by the number or amount of spin/axes a form has. eg a cube has 3 axes therefore it is dubbed 3-Dimensional. A graphic that superbly shows the 3 internally intersecting right-angled (90 degrees) planes of the Cube is shown in Fig 64 and is taken from: "The Universal One" by Walter Russell.

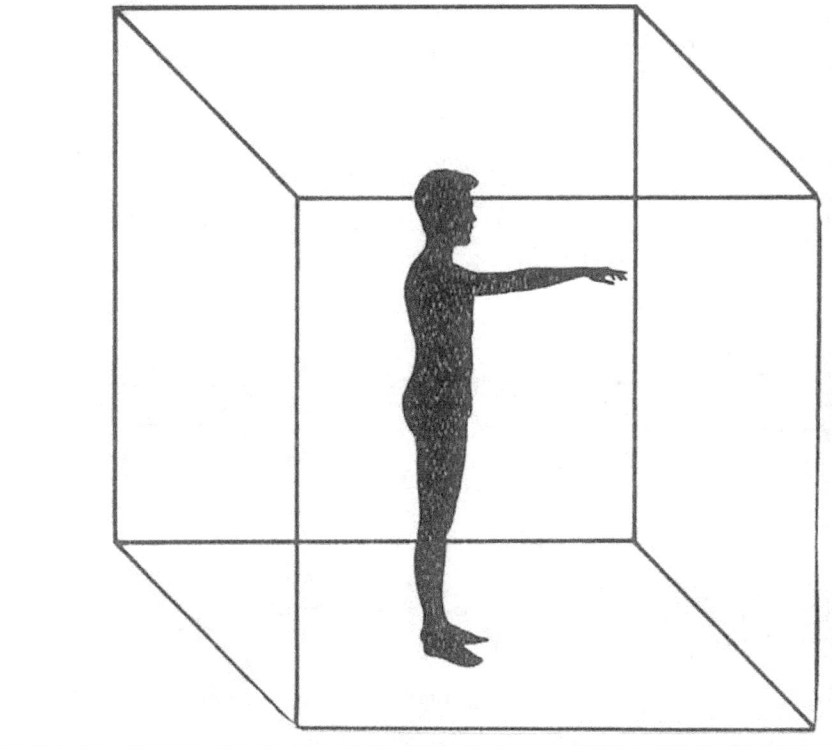

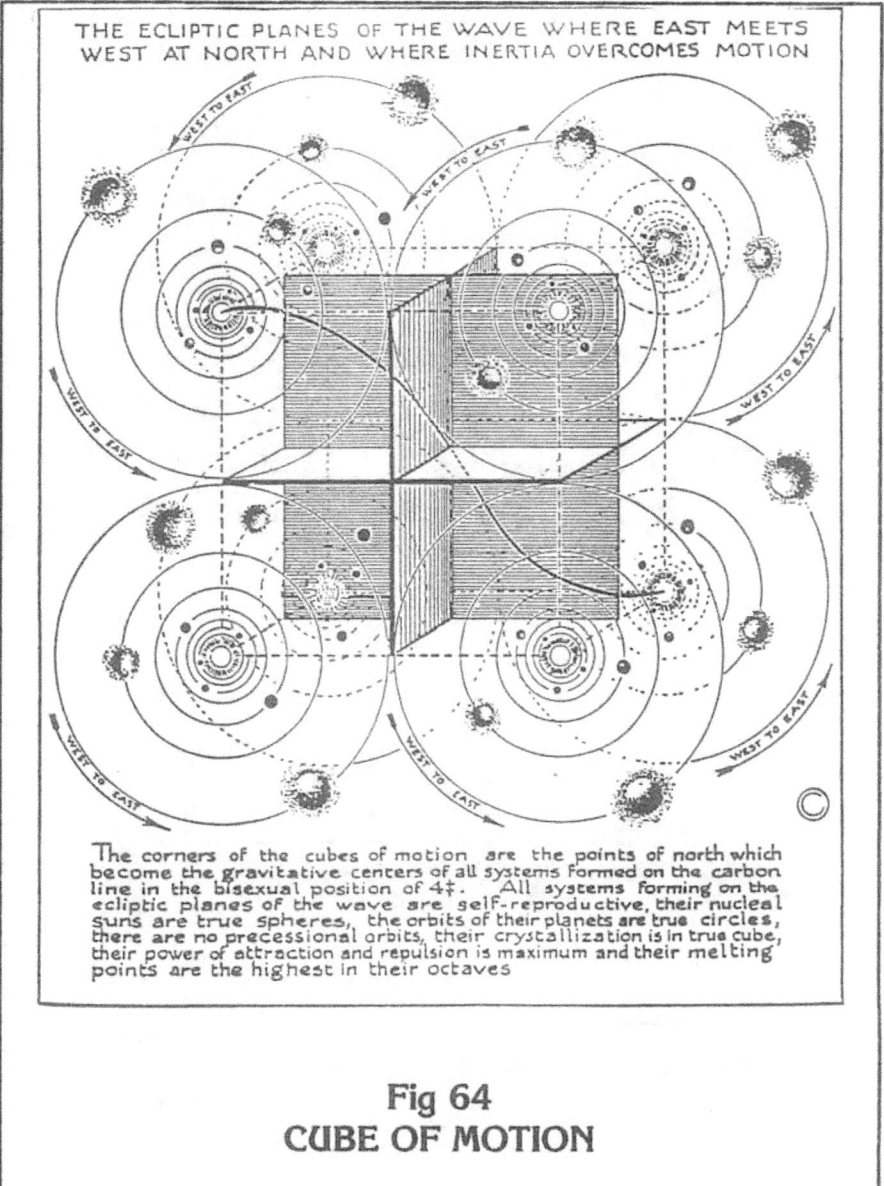

The corners of the cubes of motion are the points of north which become the gravitative centers of all systems formed on the carbon line in the bisexual position of 4‡. All systems forming on the ecliptic planes of the wave are self-reproductive, their nuclear suns are true spheres, the orbits of their planets are true circles, there are no precessional orbits, their crystallization is in true cube, their power of attraction and repulsion is maximum and their melting points are the highest in their octaves

**Fig 64
CUBE OF MOTION**

The Derivation of Phi Using Graphs

Here is an unusual and rarely seen derivation of Phi using two important graphs depicting
1): Linear Growth   y = x + 1 and
2): Exponential Growth   y = x squared.

Perhaps the most simplest graph that any student could draw, by joining dot to dot, is a graph that merely adds one unit to itself creating a Linear Graph. Adding the same amount, increasing by one unit, is a Vedic Sutra known as '<u>BY ONE MORE</u>', thus giving the name of our graph as:

y = x + 1

This is the simplest expression or function of **growth** and is the key to solving the mystery of Phi. Technically, this linear function is called an Arithmetic Proportion. (In my previous book, you learnt that Phi is both an Arithmetic and Geometric Progression).

This linear growth is shown as Fig 65

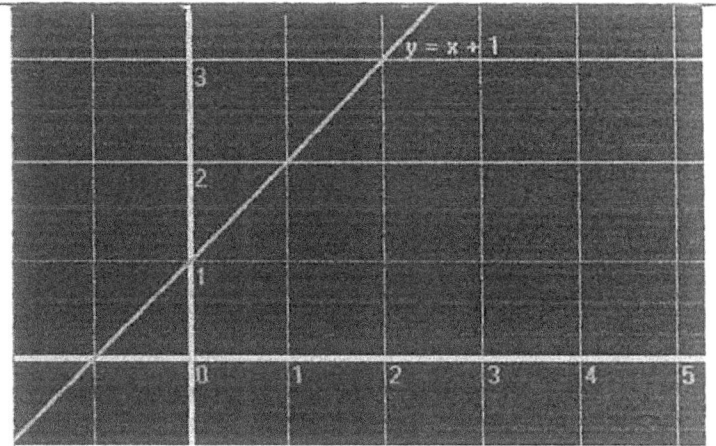

Fig 65   Linear Growth of y = x + 1

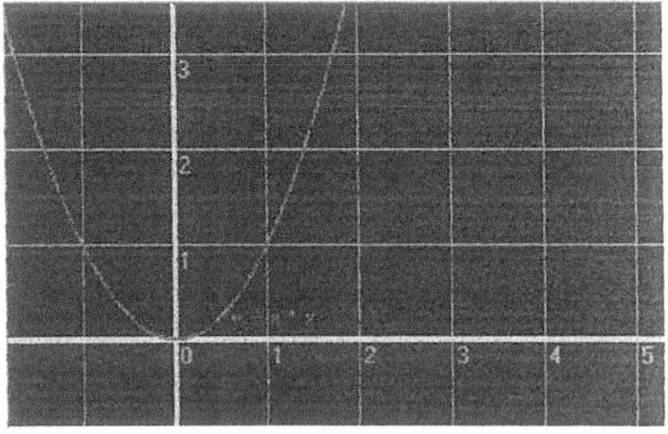

Fig 66   Exponential Growth of $y = x^2$

For the second graph, instead of adding 1 or the same relative amount, we are going to multiply this amount by itself. Again, this is the most simplest case and the curved derived, by simply joining dot to dot, is a parabola. Algebraically, we describe this as

$$y = x^2$$

It is called 'EXPONENTIAL' as the power is 2 which is called the exponent. Exponential functions are found in systems that grow like populations of humans or viruses. They can also grow very rapidly. (In this book, you will discover a new formula for "e" symbol of the exponential number 2.718 etc another transcendental number that apparently goes forever to infinity).

Such exponential functions are termed Geometric Proportions.

Study this shape shown on the previous page as Fig 66.

What would do you expect to find if we were to overlay or superimpose these two graphs, one upon the other? This is shown as Fig 67.

Can you see a Golden Rectangle, whose top right hand corner has a circle around it? Can you see some other smaller Golden Rectangles?

Observe the line directly below the small circle and observe where it hits the horizontal "x" axis.

You will have determined that the value for "x" is 1.61803398874989.

Along the vertical "y" axis, the height of this Golden Rectangle is "phi + 1" which
= 2.61803398874989.

Interestingly, this value for "phi + 1" is also the value for "phi squared" or "phi x phi".

This observation will help you to understand that **Phi knows how to add and multiply itself**, so uniquely, that there is no known other entity that behaves like this.

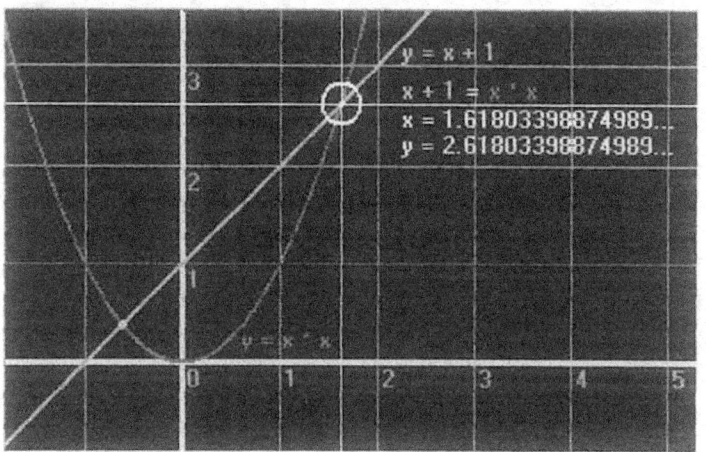

Fig 67
PHI DERIVED GRAPHICALLY
Combining y = 'x'+1 and y = 'x' squared

Knowing the Fibonacci Numbers in advance would help to predict this answer but if you look at Fig 67 again you will notice that there exists another solution, but in another dimension, in the negative realm, which is to the left of the Origin or zero point. Observe where the small circle is in Fig 68 and drop a line vertically down until it hits the horizontal "x" axis. The Golden Rectangle thus formed has its "x" axis length as "–1/phi" (minus or negative reciprocal of phi) which is:

"–1/phi" = **–.61803398874989**.

The vertical width of this smaller golden rectangle, as described on the "y" axis has a length of "–1/phi + 1" or "– 1/phi x –1/phi" (ie: minus reciprocal of phi times minus reciprocal of phi), or minus the reciprocal of phi squared, which is: – (.618) squared = **– .38196601125020**.

This graphical representation of phi could have been the first chapter of this book, as the union of $y=x+1$ and $y=x^2$ make manifest the existence of this entity called Phi. Then it follows to derive the exact value of Phi which was the first chapter of this book but done skillfully via the magic of Vedic Mathematics, which employs amazing short-cuts that bypass the clumsy quadratic equations that were taught at school.

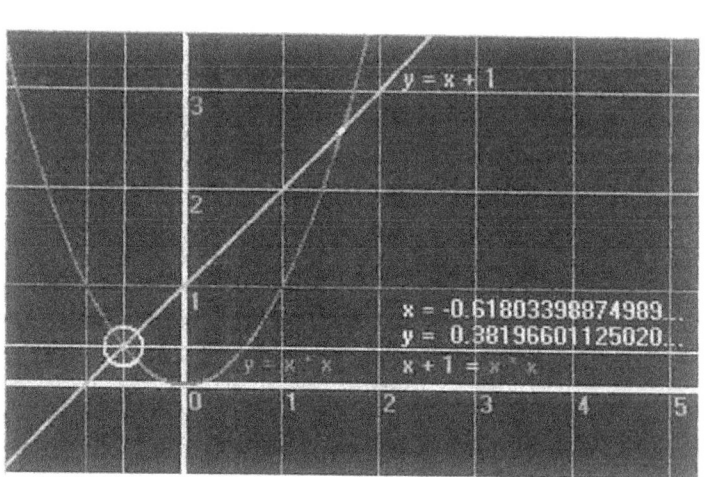

Fig 68   The Unexpected Solution of Negative Phi

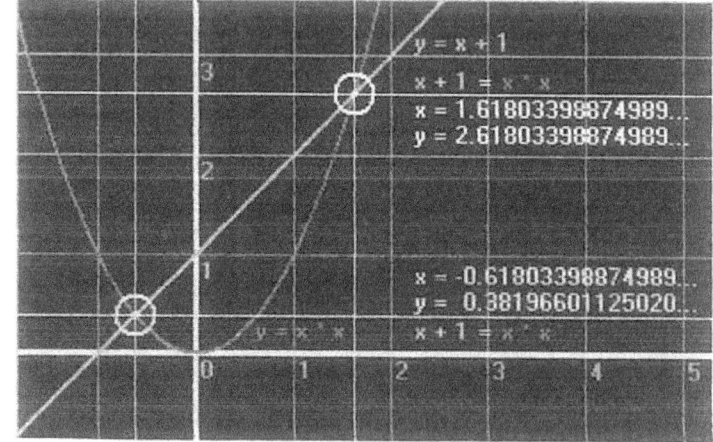

Fig 68a   Showing Both Solutions

THE PHI CODE OF 24 RECURRING DIGITS AND RODIN'S COIL

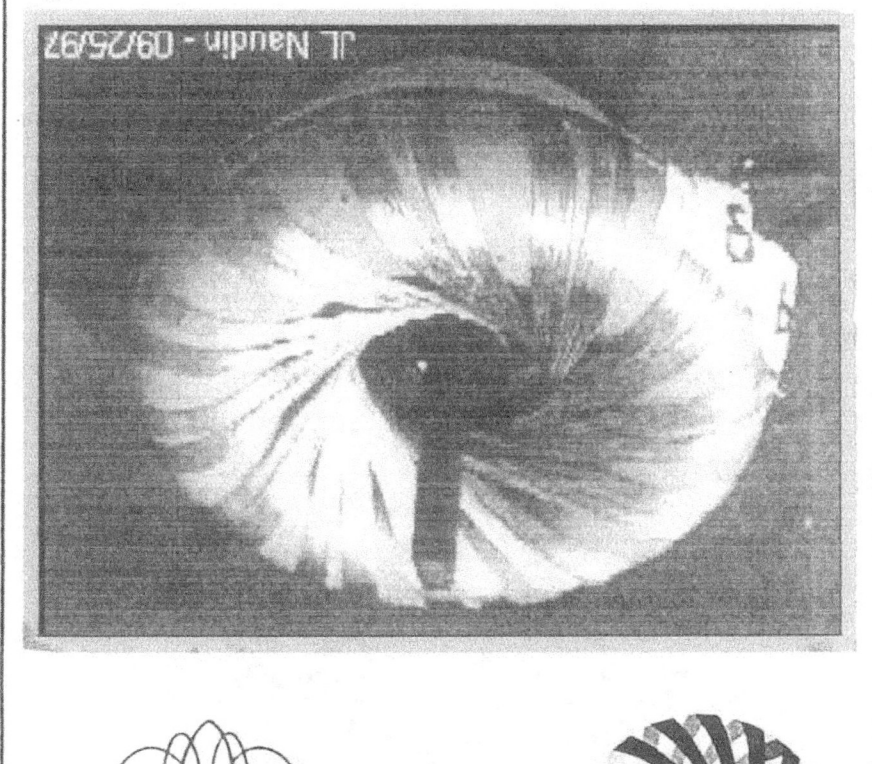

Fig 69
Marko Rodin's unusual coil design

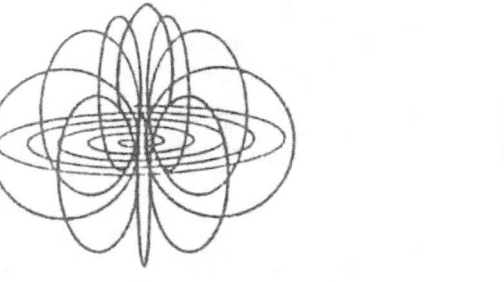

To begin this segment, we will need to understand more about the Torus (as shown in Fig 69 as Rodin's Coil and Fig 70), which is the shape of the vortex and appears widely in natural phenomena like smoke rings, magnetic fields, neutron stars. It is likened also to the Hypersphere or 4th Dimensional Sphere. Imagine a sphere with an extended equator, like the ring around Saturn, and thousands of similar spheres are placed around this equator, adding another dimension or axis, forming a donut shape. The formula for the torus is the same as the HyperSphere: $2pi^2r^3$. It is a veritable model of the universe and today many scientists are excited about this shape as it explains black holes and wormholes in space. Curiously, when a drop of ink is dropped into a vessel of pure water it forms into 7 smaller toroidal donut shapes! This is pure science.

Fig 70 shows the torus as a true symbol for Phi (Φ). The 2-Dimensional symbol, taken from the Greek Alphabet, depicts a line penetrating the circle which in 3 Dimensions is the vortex emanating from the Torus. It also appears as a fertility symbol.

Dr Derald Langham, a geneticist who coined the term "Genesa", shows how the sea-urchin embryo starts as a single cell, is then dimpled by sperm causing it to turn inside out. The cell which is a sphere dimples turns inside out into a torus, from

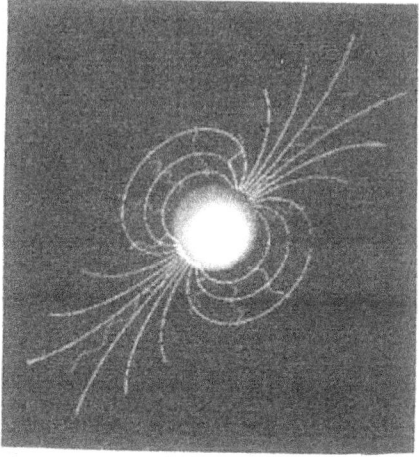

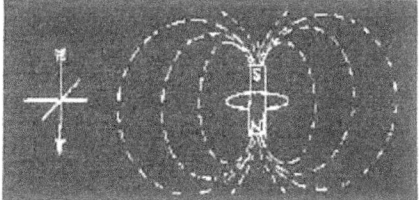

Fig 70
Torus = Symbol for Phi

the sphere to the torus, the turning inside out creates a unified field. It seems to be the only definition Science has for Consciousness, is the ability to turn inside-out. By turning inside out, we get to know ourselves, we become self-aware.

In his visionary book "Aerodynamics," Marko Rodin expresses his vision of a four-dimensional sphere which becomes a complicated toroidal structure when projected into three dimensions. He perceives a mapping of an energy flow on the surface of this projected toroid. The Rodin Coil expects to get more energy output than what is put in, in a sense, a doubling of energy. Fig 71 shows his now famous Toroid Sunflower Map which is essentially derived from numerology and rendered upon the surface of the torus. He expects strong gravitational effects from a properly executed and powered winding.

He begins with the sequence of powers of two: 1, 2, 4, 8, 16, 32, 64, 128 etc and reduces these numbers to single digits (which is essentially the Vedic Maths Sutra of Sammacaya: Digital Compression. Ancient traditions believed that the single units carry as much weight as the tens or hundreds or thousands and in a sense act like the outer atom of an electron that knows valence symmetry).

eg $64 = 6 + 4 = 10 = 1 + 0 = 1$.

This gives the reduced series of repeating numbers: 1, 2, 4, 8, 7, 5, 1, 2, 4, 8, 7, 5 etc.

Marko claims that this doubling sequence is seen in all life processes, especially in nature. The numbers 3, 6, 9, have a special place in this system, symbolising Nature's trinity and are seen in Fig 71 as

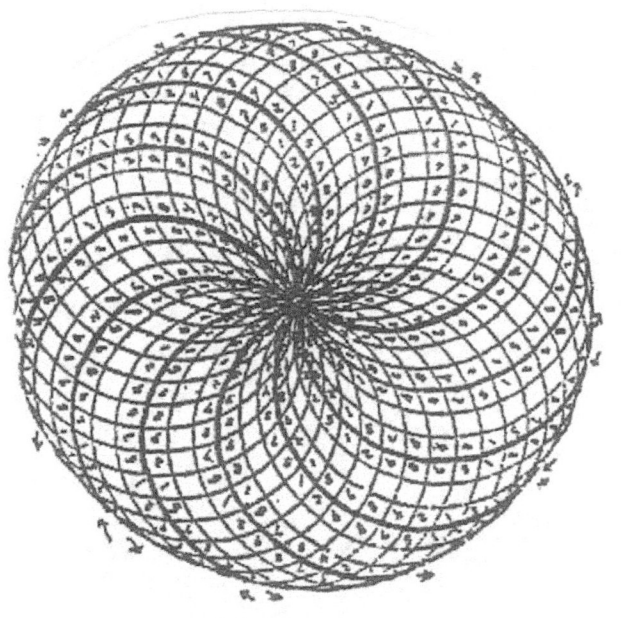

a Gap Circuit, a sequence of 3, 9, 6, 6, 3, 9, 6, 6, 9 etc. Observe also the binary sequence of 1, 2, 4, 8, 7, 5, and another doubling sequence going in the opposite direction: 5, 7, 8, 4, 2, 1.

Fig 71
Marko Rodin's Toroid Sunflower Map, showing the two doubling circuits of his polarised fractal geometry. The gap space is left blank, but contains the sequence 3, 9, 6, 6, 3, 9, 3, 3, 6, 6, 9, 3

Fig 71 will soon reveal how its dance of energy relates to the Phi Code that I have glorified as some kind of futuristic Rosetta Stone as the key to Vortex Mathematics which is the Higher Dimensional Mathematics of Spacecraft.

We know that the Fibonacci Series generates Phi and contains an explicit or distinct periodicity of 24 and that these 12 Pairs of 9 generate a CASCADE BETWEEN FREQUENCIES which is really what Phi is about. Phi is not just a number like 1.61803398 etc but a function of Nature, a Principle of the Universe that incorporates many frequencies for the expression of total health.

Fig 72 shows the numerological Fibonacci Sequence as a repeated cycle of 24 elements. Notice that the diametrically opposing numbers add up to 9. Also, the diametrically opposing lines add up to 9. The ability to do this is known as **BIPOLAR SYMMETRY.**

In Fig 73 we superimpose the linear doubling sequence of 1, 2, 4, 8, 7, 5, onto the circular expression of the Phi Code of 24 segments. Notice the hexagonal geometries formed depicting the 2 doubling sequences and the gap circuit. In fact these three hexagonal associations geometrically express three sixes (666) and the fourth hexagon is not drawn as its function still remains uncertain.

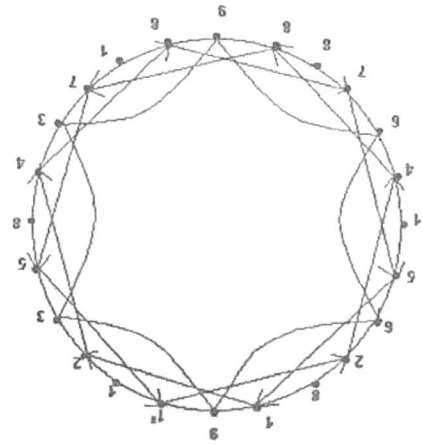

Fig 72
Phi Code of 24 Digits as a Recursive Wheel

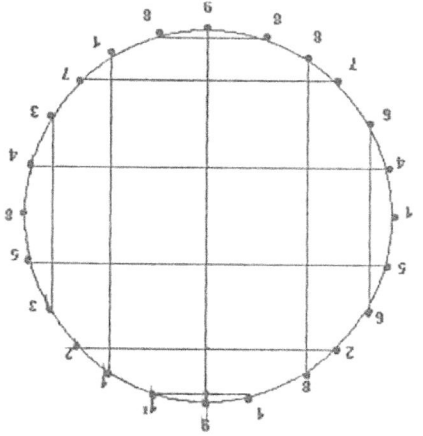

Fig 73
Two Doubling Circuits and Gap Circuit in Phi Code

If DNA is a double helix that joins end to end to form a ring or torus then these diagrams are perhaps <u>etheric templates in DNA's light filaments</u> that stores an individual's galactic memory and guides their evolution. The ability to compress numbers into single digits is not mere numerology but a sophisticated and intelligent form of number crunching or encription compression. It is no mere accident that the bilateral symmetry of the decimal system results in polar number pairs that have parity by being perfect mirrors of one another. The Number 9 appears to be an important node or bridge.

Having self-realised the importance of the Phi Code, I know that Rodin's doubling circuit .numbers 1,2,4,8,7,5 plugged into the phi code would generate a very efficient electrical coil. I don't fully understand his gap circuit of 3,6,9 suffice to say that we must think of the Yin/Yang not as a duality but rather as a trinary. (The numbers 3, 6, 9 are the numbers missing in the doubling circuit 1,2,4,8,7,5).

We need now to ask, what mathematical model contains both the doubling sequence and the fibonacci numbers. In effect what model contains both Figs 72 and 73. This would be perfect compression. The answer is found in <u>Pascal's Triangle</u> also known as <u>Halayudha's Triangle</u> aka Meru-Prastera. He was a Jaina mathematician in the 3rd century BC, and realised that the sum of any of its horizontal rows produced the Powers of 2. That the diagonal lines going through the Triangle produced the Fibonacci Numbers. He also knew that each successive line was really an increasing Power of 11. eg the 3rd line is 1, 2, 1 which is 11 squared; the 4th line is 1, 3, 3, 1 which is 11 cubed; the 5th line is 1, 4, 6, 4, 1 which is 11 to the 4th Power etc.
This is shown below as Fig 74. (For more details, see my book: THE VEDIC MATHS CURRICULUM FOR THE GLOBAL SCHOOL. Book 1: DIGITAL SUMS. Jain. 2001).

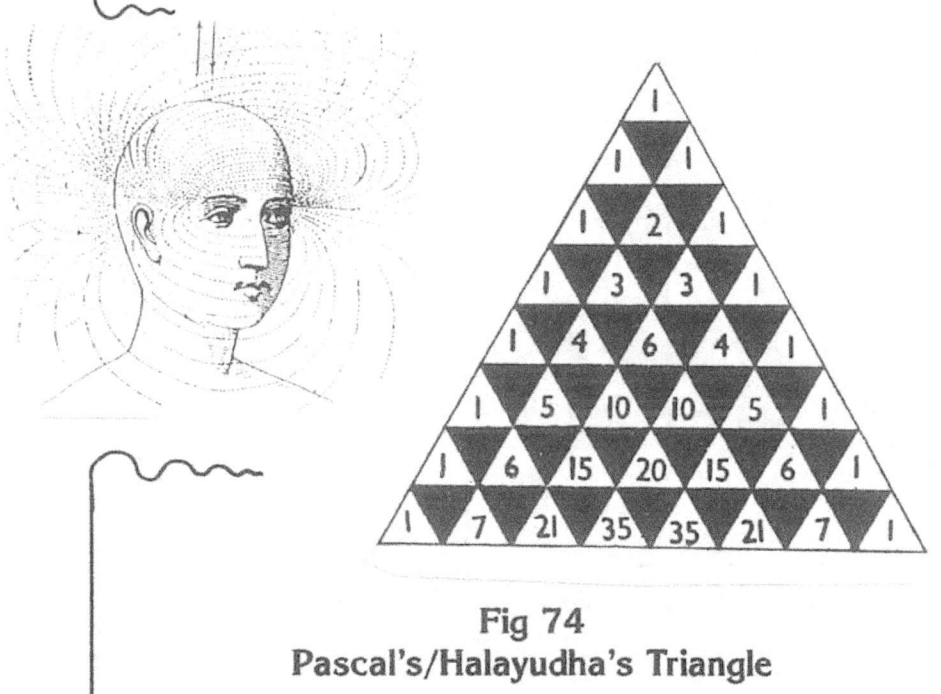

Fig 74
Pascal's/Halayudha's Triangle

Plotting the 24 Reduced Digits SUCCESSIVE DIFFERENCES.

The winding pattern that produces the synchronised electricity may be examined at a deeper level by observing the differences between the successive or consecutive order of the 24 recurring digits as shown in Fig 75.

We are not examining the doubling sequence here which orients the electrons moving through the windings. Such a winding minimises random collisions of electrons, as well as heat, reluctance and friction. The right and left doubling circuits seem to conduct the flow of electrons in opposite, parallel diagonal directions. The gap spaces, which separate the windings act like equal potential major grooves. Its been shown, with a long ferrite rod, that the centre of the Rodin Coil is quite active.

Relating only to the Phi Code, would we expect to find any internal symmetry within the differences between successive or neighbouring numbers. Fig 75 gives all the information in three columns. Column 1 shows the Fibonacci Nos. Column. 2 is the Phi Code produced by adding the digits of Col. 1 to single digits. Column 3 is the critical analysis that measures the successive differences of the Column 2 numbers either being positive (+) or negative (−). This may represent an energy gap, voltage or charge.

	Column 1	Column 2	Column 3
1.	1	1	0
2.	1	1	0
3.	2	2	+1
4.	3	3	+1
5.	5	5	+2
6.	8	8	+3
7.	13	4	−4
8.	21	3	−1
9.	34	7	+4
10.	55	1	−6
11.	89	8	+7
12.	144	9	+1
13.	233	8	−1
14.	377	8	0
15.	610	7	−1
16.	987	6	−1
17.	1,597	4	−2
18.	2,584	1	−3
19.	4,181	5	+4
20.	6,765	9	+1
21.	10,946	2	−4
22.	17,711	8	+6
23.	28,657	1	−7
24.	46,365	6	+8
	75,025	1	−8
	121,393	1	0

Fig 75 (+ve and −ve Differences of the Phi Code).

The data collected in Column 3 of Fig 75 shows some kind of symmetry but this is not clear until we graph this information, to make the hidden pattern visible. When discovered it is quite startling.

Fig 76 shows the differences between the successive Phi Code Numbers along the horizontal x-axis. This includes all the voltages from −9 to +9 (negative 9 to positive 9). On the vertical y-axis are the 24 recurring digits of the Phi Code. When we plot the information collected in Column 3 you will discover the first half of the Phi Code, that is the first 12 entries beginning from the bottom of the graph to midway, shows a pathway that is distinctly repeated in the second half of the graph but IN REVERSE or inverted or mirror-imaged in reverse!

We would expect the DNA double helix molecule to behave like this or have this complementary reverse cyclic behaviour. Such a revelation inclines a scientist to accept the notion that DNA is not merely a double helix tapering into a vague nothingness but clearly its ends are joined to form ring shapes or toroids.

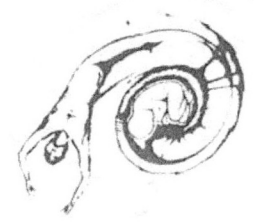

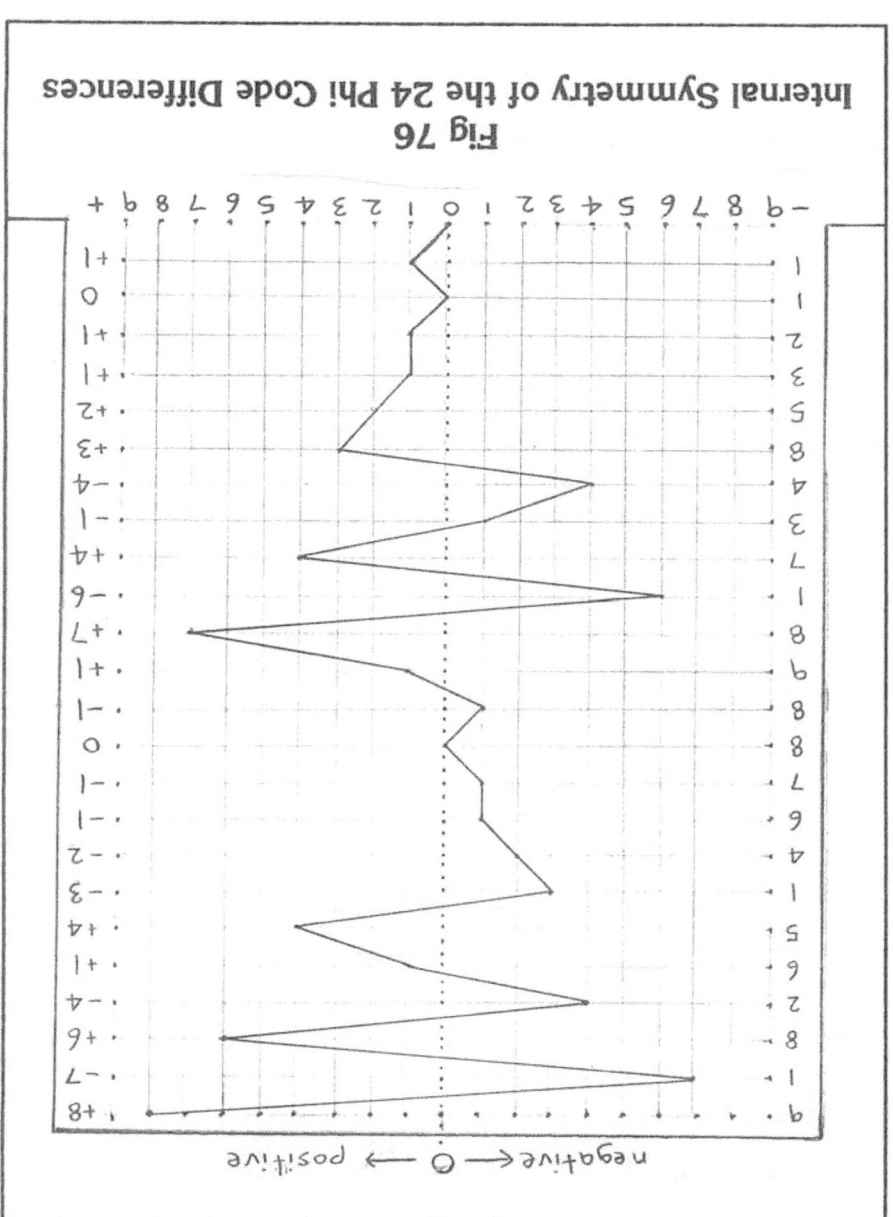

Fig 76
Internal Symmetry of the 24 Phi Code Differences

A few more graphics will help to tie together and conclude this chapter on the Rodin Coil. Many mystics and scientists agree that the two most important number systems, besides our decimal Base 10, appear to be Base 12 and Base 9. Its quite interesting that the Phi Code is really a phenomena of 12 Pairs of 9. These numbers: 9, 10 and 12 appear again in the Sino-Tibetan Zodiac of Animals that shows the Magic Square of 3 x 3 at its Centre. Refer to Fig 77.

The 9 numbers of the Magic Square are thus arranged so that all columns and rows and diagonals have a sum of 15. As Above, So Below. Equality in all Dimensions. Order amongst Chaos. You will notice also that around the central number 5 (PHI-ve) all the Pairs balanced around it have a sum of 10, as in 1 and 9, 2 and 8, 3 and 7, 4 and 6.

The 8 Trigrams surrounding this, when squared, give **64** DNA codon permutations which links in with Rodin's doubling sequence of 1, 2, 4, 8, 16, 32, 64 etc. Around this are the 12 constellations of Earth's ecliptic path. Often, this whole diagram is shown on the underbelly of a tortoise, where the numbers of the Magic Square (known as the **LO-SHU**, circa C17th BC) first appeared on Earth, representing the incarnation of the fiery God Manjushri whose role it is to protect this Knowledge.

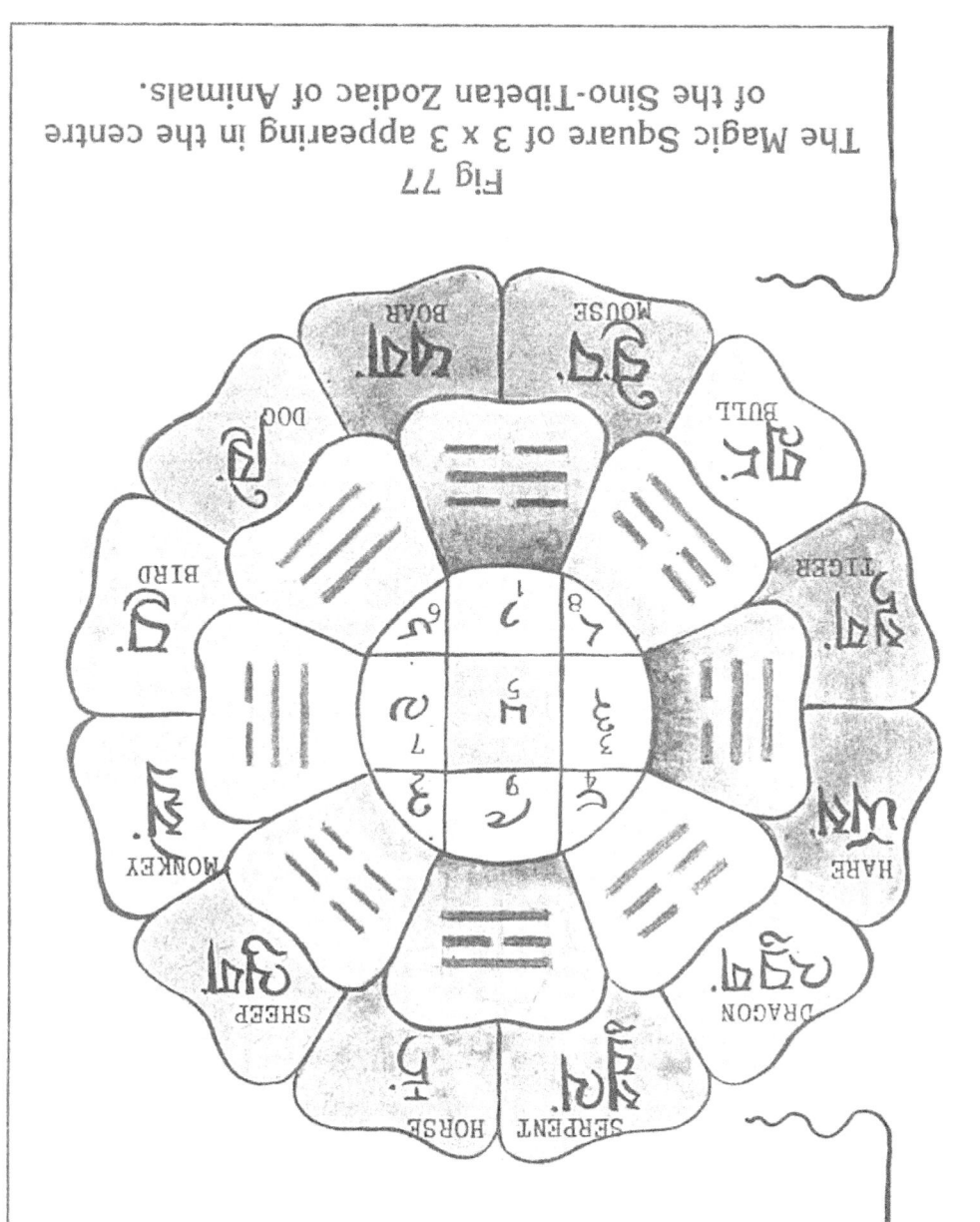

Fig 77
The Magic Square of 3 x 3 appearing in the centre of the Sino-Tibetan Zodiac of Animals.

Finally, this work is being taken seriously by a progressive French scientist Jean-Louis Naudin (www.jlnlabs.org) who has taken the intuitive work of Marko Rodin and implemented aspects of the Phi Code into Free-Energy solutions, space-propulsion systems using antigravity and propellantless propulsion. Naudin claims to use a special toroidal coil named the The Rodin Coil (see Fig 69) to induce a "twist" or a B-Field torsion effect in the Aether. This is shown in Fig 78. This picture shows the extensive magnetic field lines confirming that the Phi Code inducing the B-Field torsion effect is very real.

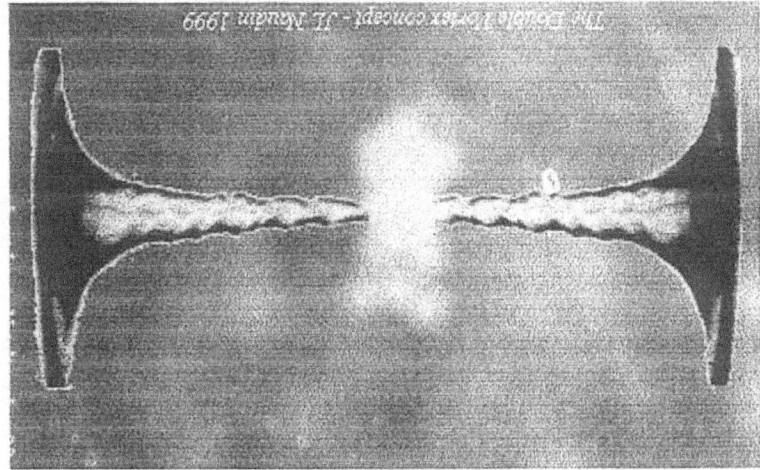

The basic principle of the double vortex generator

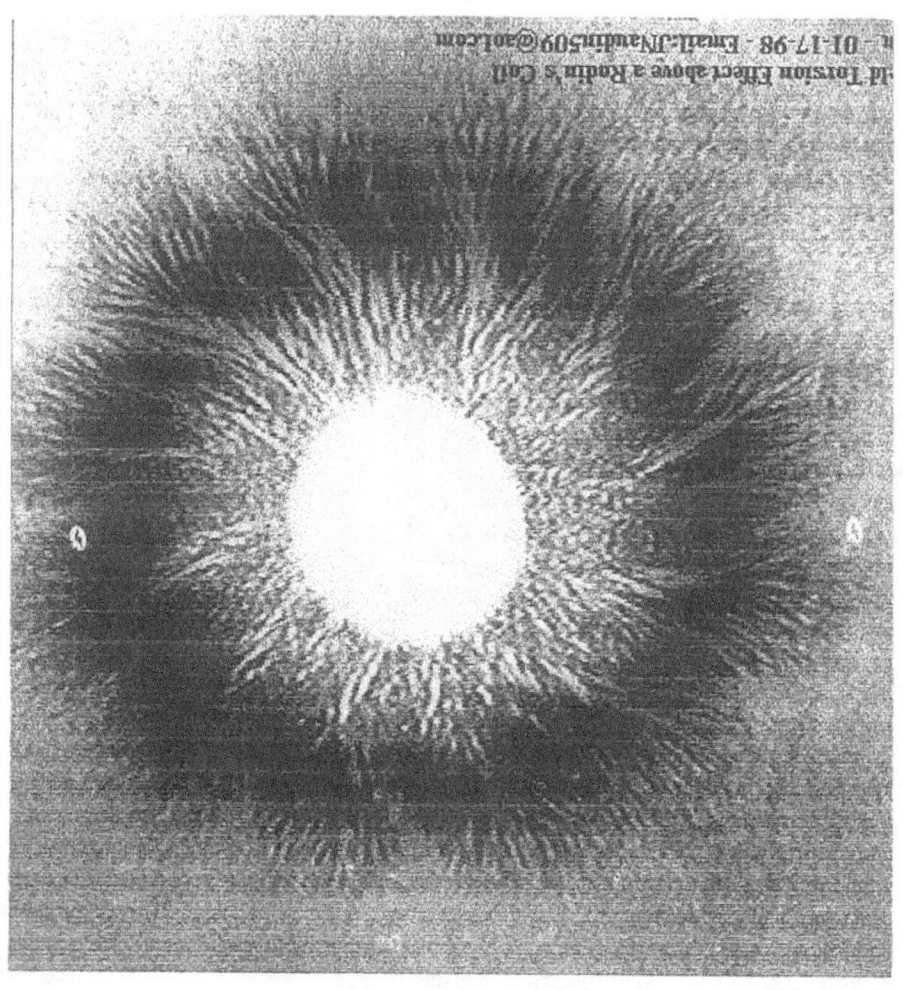

Fig 78
Phi Code Generating Extended Magnetic Fields.

VORTEX IN A BOTTLE

To understand Naudin's system of Electro Hydro Dynamics here is an project you can do at home. To get a feel for this vortex, which is an open system having differential pressures existing between two layers, Naudin suggests the **Vortex in the Bottle** experiment.
See Fig 79.
This leads on to his concept of the Double Vortex.

Hopefully this will give you insight into the essential components in the Quest for **Over-Unity**:

Vortex Mathematics based on **Phi**.

Get two plastic bottles prefer ably from a 2 litre size soft-drink bottle. Glue the two plastic caps and then make a 9mm hole in the center. Fill up one bottle 3/4 full with water. Connect the caps and bottles into one joined unit. Give a twist to this device, or merely turn it upside down with a little shake, and you will clearly see the Vortex in the Bottle. (Some people remove the caps completely and tape the two open ends).

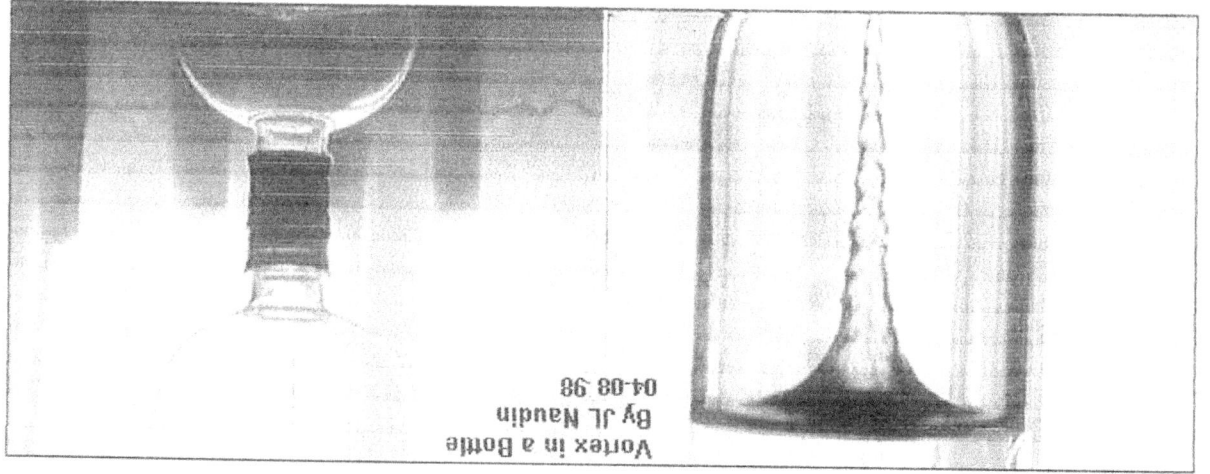

Vortex in a Bottle
By JL Naudin
04-08-98

Fig 79
Vortex in a Bottle

The concept of the Double Vortex inspired me to inspect Fig 76 again and X-ray it a bit deeper. Is there a Double Vortex in the Phi Code Differences? I believe there is, within the 24 recurring Phi Digit Differences a doubling of the essential pattern of Fig 76 occurs. This is achieved by doubling the image of Fig 76, by superimposing upon itself its own mirror image. If you had Fig 76 photocopied twice onto clear transparency and flipped one over the other you would end up with Fig 80. Another way of getting the same result is to draw Fig 76 again and plot upon it the same information but with reversed signs of polarity, eg where it has say +3, plot also −3, and where there is say a −2, plot also a +2 and you will arrive at Fig 80.

Notice how there appears to be one tornado above another tornado, both being the same size. They also appear to be like Benjamin Franklin's Magic Squares of 8 x 8 or 16 x 16 which you can study in my other books: THE BOOK OF MAGIC SQUARES Volumes 2 and 3 respectively, by Jain.

In Sacred Geometry we are continually looking for symmetry within symmetry or the nesting of such symmetries as an indication of Order amongst Chaos. This result was expected as the sequence of 24 recurring digits collapses effectively into 12 Pairs whose sums add up to 9.

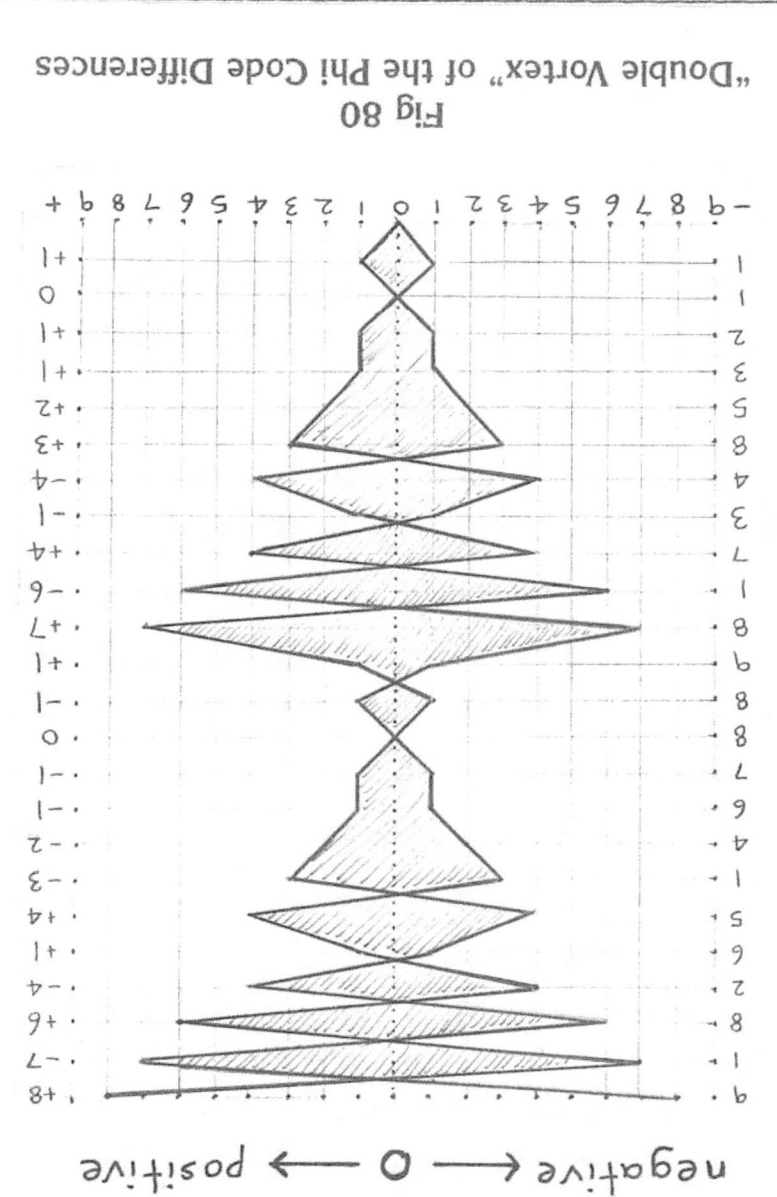

Fig 80
"Double Vortex" of the Phi Code Differences

This bipolar symmetry is also evident in the diagram of Fig 81 that shows the original 24 Recurring Digits of the Phi Code plotted circularly into the 9-Point Circle.

(For detailed information on the 9-Point Circle and why it is important, refer to my two books: THE MAGIC OF NINE IN VEDIC MATHEMATICS and THE VEDIC MATHS CURRICULUM FOR THE GLOBAL SCHOOL, Part 1: DIGITAL SUMS, by Jain).

Begin with a nonagon and isolate only the 9 vertices and have in front of you the Phi Code of 24 Recurring Digits, (Set 2 that does not contain the initial zero and has its last pair as the double 9 bond):

1, 1, 2, 3, 5, 8, 4, 3, 7, 1, 8, 9,
8, 8, 7, 6, 4, 1, 5, 6, 2, 8, 1, 9

Plot this in order as they appear, sometimes drawing over existing lines, and you will arrive at Fig 81. Notice that the pattern on the left half of the circle is identical but mirror-imaged to match the right-half of the circle. If no symmetry was formed in Fig 81 then this book would not have manifested. The continual discovery of symmetry within symmetry of the Divine Proportion warrants this information out that it may be received into the heart and mind of those who are receptive and inventive and can apply this Code to a useful and spiritual technology.

Musical notes attributed to these 9 Pairs can form celestial music.

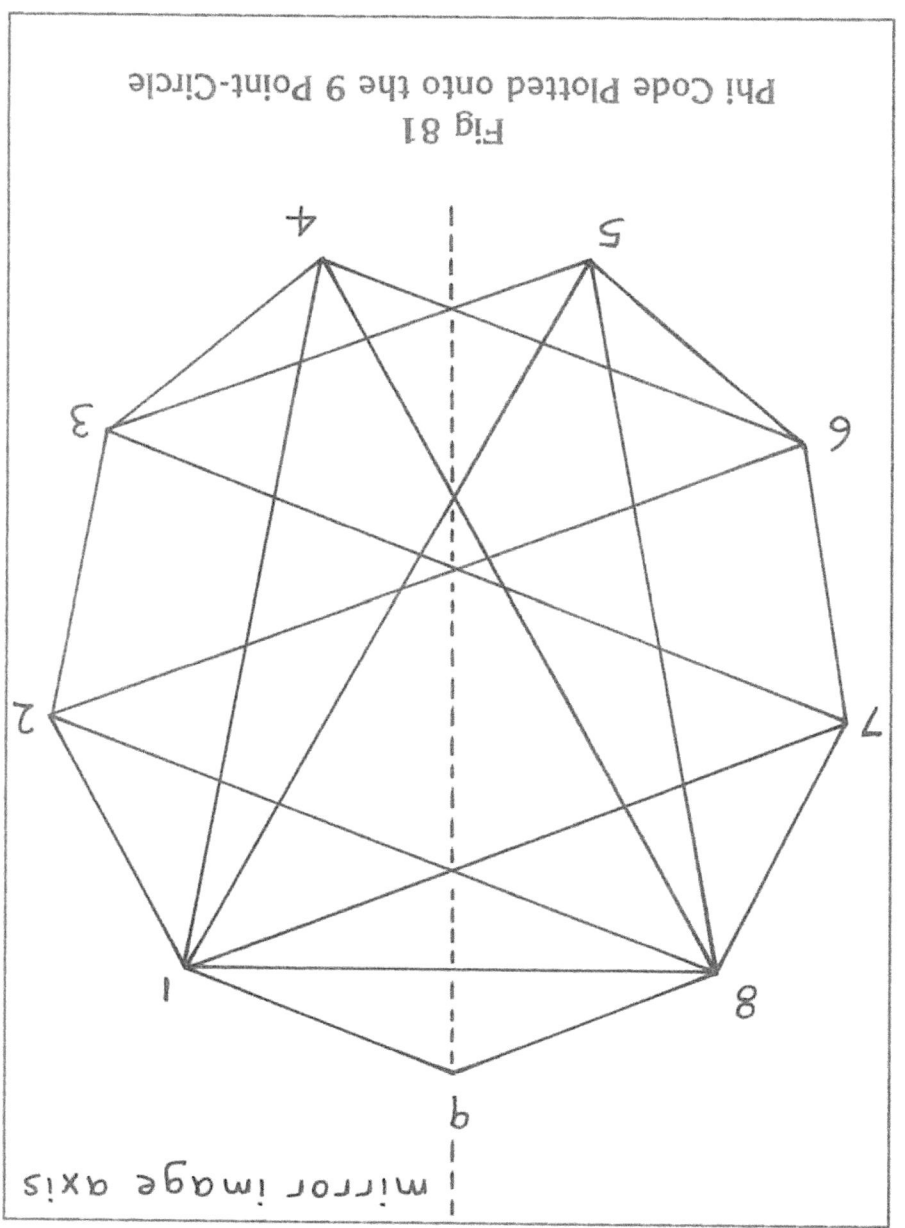

Fig 81
Phi Code Plotted onto the 9 Point-Circle

REFERENCES TO "24" IN THE 4TH DIMENSION

We know of the Platonic and Archidmedean Solids, but little of what are known as the Hyper Solids or Hyper Polyhedra in the 4th Dimension. eg the one that corresponds to the Dodecahedron is known as the Hecatonicosahedroid or C120 which has 120 Dodecahedra as 3-Dimensional bounding cells, 720 pentagonal faces, 1200 sides and 600 vertices. These obviously fit into the realm of the Golden Mean in 4th Dimensional space.

The one of interest here is C24, the **Icosatetrahedroid** which has **24** 3-Dimensional bounding cells. This was used by Claude Bragdon, the famous American architect who used 2-D and 3-D projections of the 6 known hypersolids. He had a particular penchant for magic squares and used their derived patterns ornamentally in the ceiling of the Chamber of Commerce, Rochester, New York. He also wrote a book on the 4th Dimension.

Fig 82 shows one of his illustrations of projective ornament. The statue of a Being holding a 3-D spherical object has the encaption "Geometry".

Fig 82
C. Bragdon's Illustration of Projective Ornament

Another important occurence in the 4th Dimension appears in what is known as

The Maximal Kissing Number of n-Dimensional Laminated Lattices:

0, 2, 6, 12, **24**, 40, 72,, 126, 240, 272 etc

It has to do with a Sequence of Numbers that explain Sphere-Packing in the Many Dimensions.. If we begin with a hexagon in 2-Dimensions, it is obvious that 6 spheres can touch in this shape. When we go to the 3-Dimensions, as shown clearly in Figs 24a and 24b there are 12 spheres around a central sphere, the 12 Disciples orbitting the central Christ Principle. And in the 4-D there are 24 Spheres or Hyper-Spheres surrounding this 12 Sphere Cuboctahedron. Derald Langam, in his work on agricultural genetics called "Genesa" concludes that the Cuboctahedron represents the **3rd zygotic stage of human cell-division.**

Laminated Lattices are best visualised as crystal packings. The next number in this Sequence is 40 which relates therefore to the 5th Dimension.

There is a fantastic On-Line Encyclopedia of Integer Sequences which is credited to the work of N.J.A. Sloane, J.H. Conway and other colleagues. On there, you will discover Sequences like this one and thousands of others.

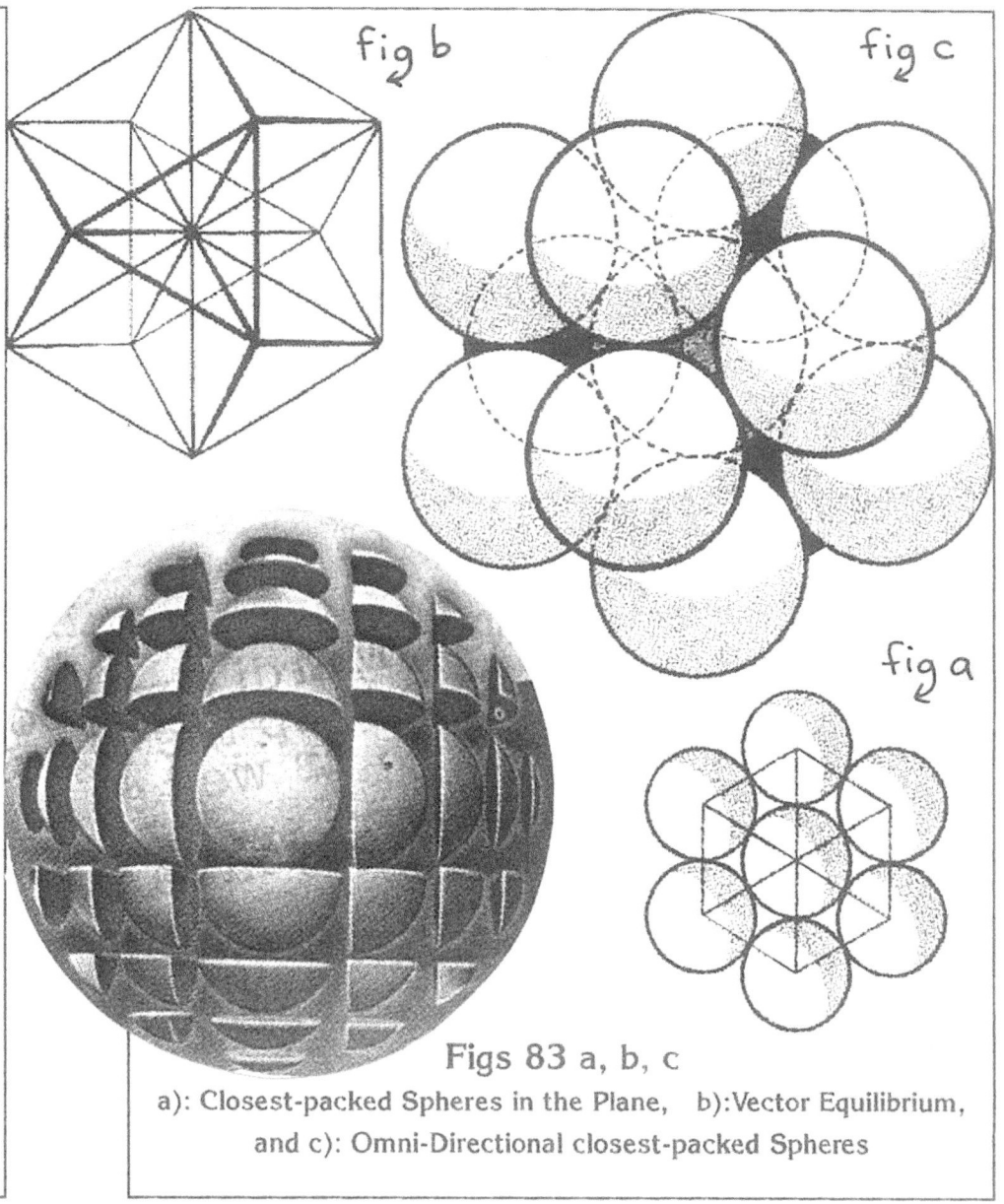

Figs 83 a, b, c
a): Closest-packed Spheres in the Plane, b): Vector Equilibrium, and c): Omni-Directional closest-packed Spheres

Eddington, on the 4th Dimension:

However successful the theory of a four-dimensional world may be, it is difficult to ignore a voice inside of us which whispers: "At the back of your mind, you know that a fourth dimension is all nonsense". I fancy that voice must often have had a busy time in the past history of physics. What nonsense to say that this solid table on which I am writing is a collection of electrons moving with prodigious speed in empty spaces, which relatively to electronic dimensions are as wide as the spaces between planets in the solar system! What nonsense to say that the thin air is trying to crush my body with a load of 14 lbs. to the square inch! What nonsense that the star cluster which I see through the telescope, obviously there *now*, is a glimpse into a past age 50,000 years ago! Let us not be beguiled by this voice. It is discredited...

We have found a strange footprint on the shores of the unknown. We have devised profound theories, one after another to account for its origin. At last, we have succeeded in reconstructing the creature that made the footprint. And lo! It is our own.

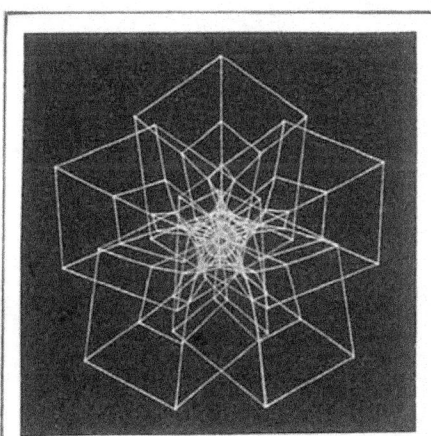

Fig 83 d

When scientists try to represent fourth-dimensional hyperspace on two-dimensional surfaces, they get exotic images:

To conclude this excursion into the 4th Dimension, have a read of this excerpt from Edwin A. Abbott's highly recommended: "FLATLAND", written more than a century ago. Narrated by A. Square, "Flatland" is a delightful mathematical fantasy about life in a two-dimensional world. Very witty, it is also a parody of Victorian society.

STRANGER IN FLATLAND

How would a three-dimensional sphere explain itself to a two-dimensional square? It wouldn't be easy, as the following excerpt from Edwin Abbott's classic novel Flatland *illustrates.*

Narrator (a Square): In what direction is the Third Dimension, unknown to me?

Sphere: I came from it. It is up above and down below.

N: My Lord means seemingly that it is Northward and Southward.

S: I mean nothing of the kind. I mean a direction in which you cannot look, because you have no eye in your side.

N: Pardon me, my Lord, a moment's inspection will convince your Lordship that I have a perfect luminary at the juncture of two of my sides.

S: Yes: but in order to see into Space you ought to have an eye, not on your Perimeter, but on your side, that is, on what you would probably call your inside; but we in Spaceland should call it your side.

N: An eye in my inside! An eye in my stomach! Your Lordship jests.

S: I am in no jesting humour. But the diminished brightness of your eye indicates incredulity. Now prepare to receive proof positive of the truth of my assertions. You cannot indeed see more than one of my sections, or Circles, at a time; for you have no power to raise your eye out of the plane of Flatland; but you can at least see that, as I rise in Space, so my sections become smaller. See now, I will rise; my Circle will become smaller and smaller until it finally vanishes.

[The Narrator takes over the story.]

There was no "rising" that I could see; but he diminished and finally vanished. From the depths of nowhere came forth a hollow voice: "Am I quite gone? Are you convinced now? Well, now I will gradually return to Flatland and you shall see my section become larger and larger."

When he regained his original size, he heaved a deep sigh; for he perceived by my silence that I had altogether failed to comprehend him. And indeed I was now inclining to the belief that he must be no Circle at all, but some extremely clever juggler.

Fig 84

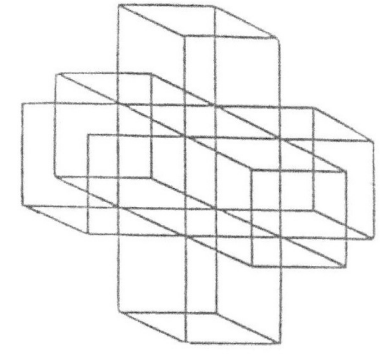

6 Cubes packed around a central cube making the image appear 4-Dimensional.

This 2-D hexagonal "Flower Of Life" Pattern is really a Cube.
Ibis-headedThoth, builder of the Pyramids in Egypt from 60,000 years ago, stands nearby.

Part 3

Two Important Original Mathematical Discoveries by Jain:

PART 3A

How the MAGIC SQUARE CONSTANTS are determined by the TETRAHEDRON!

Or How the SUM OF THREE CONSECUTIVE TETRAHEDRAL NUMBERS 1, 4, 10, 20, 35, 56, 84, 120 creates the MAGIC SUMS of the MAGIC SQUARE SERIES: 15, 34, 65, 111, 175, 260, 369

And

PART 3B

A better way to express the formula for "e" the EXPONENTIAL NUMBER

e = 2.718281828459045233

Part 3a

"IN THE NEXT DIMENSION" IS THE NAME FOR THE 17th VEDIC MATHEMATICS SUTRA.

(WE HAVE BEEN GIVEN 16 BY THE LATE PURI SHANKACHARYA OF INDIA IN 1960).

THIS IS THE LOST OR MISSING SUTRA:

THE LINKING OF THE MAGIC SQUARE CONSTANTS TO THE TRIPLETS OF THE SPHERE-PACKING NUMBERS OF THE TETRAHEDRON

This is a journey from 2-Dimensional Mathematical Sequences that have the intelligence to cross the dimensions. We do it everyday. If we have a problem in this 3-Dimensional World, we go to sleep, enter the 4-th Dimension, Dream, and come back knowing how to solve our problem. One of the main Vedic Maths Sutras is "By One More Than The Previous Digit" which is often shortened to "By One More". In this regard, we can state: "By One More Dimension".

Thus, this is <u>the Lost Sutra</u>, knowing how to simply and intelligently solve problems "<u>IN THE NEXT DIMENSION</u>". Its about knowing how spheres prefer to pack as a Tetrahedron rather than laying flat as 4 spheres on the ground. 8 spheres prefer to compact as a Cube, and 12 spheres prefer to enclose a central nucleus forming the Cuboctahedron.

Remember in the Movie "CONTACT" with Jodi Foster, where she was trying to decode galactic messages or impulses from outer space that came cryptically in the form of the Prime Numbers (numbers than can not be divided by any other number except for 1 for itself). Scientists were trying to solve this problem of interpreting all this data on their sheets and it was eventually shown to them that if the sheets were arrayed 3-Dimensionally in the form of a Tetrahedron, then and only then could the data be read.

This is a good example of how problems can be solved when we step out of the linear approach and risk a move In The Next Dimension.

Thus such a discourse leads inevitably into the mysterious realms of the Platonic and Archimedean Solids, and further.

(Please take the time to inspect again Figs 48 to 52 which show the translation of Magic Squares into Sacred Geometrical Symbols, literally the Magic Square of 3 x 3 creating King Solomon's Seal.

Also view Figs 58 and 59 showing the translation of Magic Squares into 4th Dimensional Objects, literally, the Magic Square of 4 x 4 rolled into a cylinder to create the Torus or 4-D Sphere).

To help you understand the importance of the forthcoming Magic Square Revelation hinted in the title and how it relates to solving problems In The Next Dimension, I will first take the liberty to define:
1. What are the **Magic Square Constants?**
2. What are the **Squared Numbers or Powers of 2?**
3. What are the **Triangular Numbers?**
4. What are the **Tetrahedral Numbers?**
5. What are the **Square Pyramid Numbers?**

Then you will be able to start reading a new language, the Language of Symbols, of Fixed Eternal Timeless Design. You could look at the following diagram and know exactly what it means:

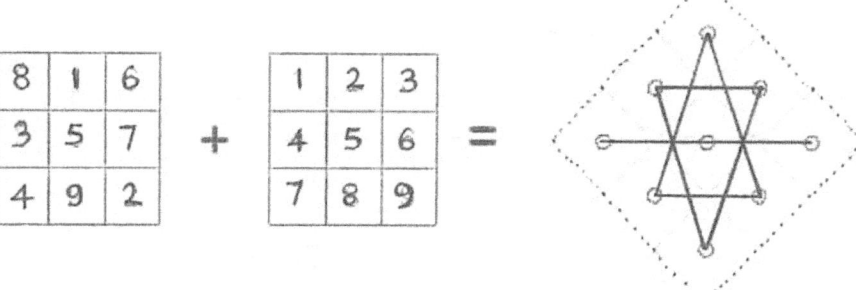

Fig 51 In a Revised Universal Form Star of Solomon Formed from the Lo-Shu

Part 3a   Section 1
WHAT ARE THE MAGIC SQUARE CONSTANTS?

Here are the 7 traditional Magical Squares, in Fig 85, those curious numerical patterns whose gift is "AS ABOVE, SO BELOW" whose top row equals its bottom row or any row or any column or any diagonal are equal. Equality in all Dimensions. As celestial archetypes, they have exercised monumental influences on thinking people who, for thousands of years, have attempted to decipher a hidden intelligence that they appear to betray.

Magical Squares are also known as Planetary Squares since the time of the Babylonians, as each configuration corresponds to a specific planet or heavenly sphere with specific Divinities or Intelligences that preside over them. Their symbols are shown at the top of each square, in Fig 85. Notice also how the square becomes larger, the closer one comes to the earth. eg Saturn is the furthest and its number is 3 x 3, whilst the Moon is the closest and its number is 9 x 9.

Fig 85
The 7 Traditional Magic Squares

What is of importance here is the harmonic or Sum of each of these Squares. This **Magic Sum** is also known as **THE MAGIC SQUARE CONSTANT**. In the chart below, let M.Sq.C represent the Magic Square Constant and "n" represent the Magnitude or size of the Magic Square. We often call a Magic Square of 3 x 3 as n = 3.

MAGNITUDE of the M.Sq.	M.Sq.C or CONSTANT	Planetary Meaning
for n = 3	C = 15	Saturn
for n = 4	C = 34	Jupiter
for n = 5	C = 65	Mars
for n = 6	C = 111	Sun
for n = 7	C = 175	Venus
for n = 8	C = 260	Mercury
for n = 9	C = 369	Moon
for n = 10	C = 505	Earth

If you wanted to know what the Constant is for an 11 x 11 M. Sq. you would plug n = 11 into the universal formula:

$$C = [n.(1 + n^2)] \div 2$$

This gives a Constant of C = half of [11.(1 + 11x11)]
nb: the '.' period = multiplication.
C = 671

There is another way of achieving the M. Sq. Constant, geometrically. Its important to know this, as most Maths Students are visual rather than logical, ie: right-brain rather then left brain. The right brain is creative, feminine, visual, intuitive, holographic. The alternate geometric method relates to a Vedic Maths Sutra called "Vertically and Crosswise" which employs the "X" shape, hinting at the cross-overs of the optical nerves. By colouring in lightly all the numbers of the diagonals of any Natural Square (Fig 86) and adding their sum gives the Magic Square Constant. Its that simple. Below, the diagonals of the first 9 numbers in arithmetic order both add to **15**:

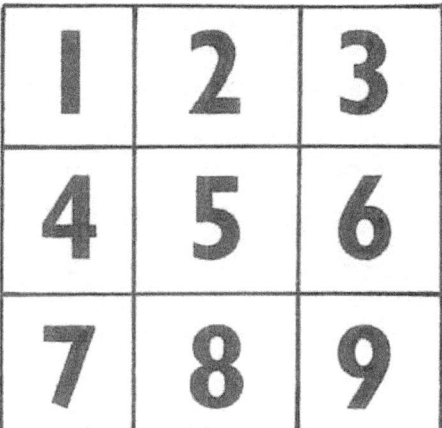

Fig 86
The Diagonals of The Natural Square of 3 x 3
Determines the Sum or Constant of the Magic Square.

Now that you understand that there are alternative geometrical solutions to common logical problems, <u>can we arrive at the same series of Magic Square Constants by other geometrical means</u>?

In fact there are two more: Methods A and B.

<u>Method A</u>:

The first is the title of this chapter, that each Magic Square Constant is the sum of 3 Tetrahedral Numbers. (We have not defined the nature of these tetrahedral numbers yet, but they are listed below. There will be a discourse on the tetrahedral numbers coming soon in this chapter 3A, Section 4).

M. Sq. Constant Series = 15, 34, 65, 111, 175, 260 etc and the Tetrahedral Numbers Series =
1, 4, 10, 20, 35, 56, 84, 120, 165, 220 etc

It occurs naturally in Nature that:

1	+ 4	+ 10	= 15	the M. Sq. C for 3 x 3
4	+ 10	+ 20	= 34	......4 x 4
10	+ 20	+ 35	= 65	......5 x 5
20	+ 35	+ 56	= 111	......6 x 6
35	+ 56	+ 84	= 175	......7 x 7
56	+ 84	+ 120	= 260	......8 x 8

The purpose of this discovery is to elucidate how all the knowledge of my last 24 years of research on Magic Squares can be encapsulated into the SHAPE of the TETRAHEDRON, which is the Male aspect of the building block of all creation whether it be in the atomic or galactic realm. Conversely, the Feminine Aspect of Sacred Geometry is the Golden Mean Spiral or Phi-Spiral, the Living Maths of Nature. All curved lines are Feminine, all straight lines are Male. When we marry the two worlds, the invisible <u>LANGUAGE OF LIGHT</u> is made visible. <u>SHAPE stores MEMORY</u>. Thus these mere curious Magic Squares are intimately linked to our own <u>Tetrahedral Consciousness</u> and is thus a storehouse, a veritable repository of the vibrational, <u>AKASHIC RECORDS</u>: the living library or sum total of all physical, emotional, intellectual and spiritual experiences. Thus in a sense, if you can remember the Numbers of the Lo-Shu or Magic Square of 3 x 3, linearly as in:
6, 1, 8, 7, 5, 3, 2, 9, 4 you can regard it as a communication/telephone-to-God device or tetrahedral code into the Higher Realms of Knowledge.

It is for this reason that I have created Mandala Colouring Books for Children. A child may think that they are creating and colouring in beautiful Magic Squares but really they are subliminally accessing its encoded memory that they already know. Its really about <u>Remembering</u> this Sacred Geometry. It was only through our separation from God that we forgot who we are. Say to yourself: I am at One with all Knowledge, all Love. **I AM THE SUTRA**. I AM PHI.

Method B:

The following diagram, Fig 86, is another geometric discovery/revelation of the Magic Square Constants. It is known in the Maths World as the Geometric or Polygonal Number Series. It charts all the family of numbers from Triangular to ELEVEN–agonal and with a stroke of genius, highlights (in bold) a descending diagonal of numbers, which is The Magic Square Constants, in order:
1, 5, 15, 34, 65, 111, and **175** and **260**!
(nb: the first two numbers: 1 and 5 are for n=1 and n=2. There are no magic squares of 1 x 1 and 2 x 2 but they are listed as part of the series).

Sometimes new scientific or mathematical discoveries are received by several people at the same time. It is one of those experiences where you think you have independently realised a great truth, only later to discover that several other authors (like R. Brooks, 2001) have it already in print and perhaps we all received it at the same time. Simultaneous transmissions. In search of patterns within patterns. If the Magic Square Constants Series is important then it must appear in other forms or disguises. This is one of the central themes of this book that **the Student of Geometrical Enquiry is a Pattern Hunter**. Whether you are a 10 year old student, a Mathematical Monk or a Numerical Nomad, the most conducive way to teach mathematics and make it fun is through the Art of Discovery. This sets up a 360 degrees view akin to the intuitive state of "**gnosis**" or knowing. (This comes from the Greek word for spiritual knowledge).

My gnostic Revelation about the Magic Square Constants revealed as Triplets of Consecutive Tetrahedral Numbers has not appeared in print, to my knowledge, . If knowledge is of any value is must be **SHAREABLE**, free, eco-sophical and it can never be owned or copyrighted.

You may look at Fig 86 and be daunted by the true meaning of say a Hexagonal Number. So I include another important diagram, Fig 87, that gives you a picture of the all the Polygonal Numbers up to Octagonal.

The implications of Fig 86 are far-reaching. What we have here is a tool to predict the next series for Twelve-agons. The secret to do this is to know that the differences between all the numbers in the vertical columns is a Triangular Number (ie 1, 1+2, 1+2+3, 1+2+3+4 etc). Observe in Fig 86 the top row of numbers that reads: (Differences in the Vertical Values). Again, via **Pattern Recognition, we are able to Predict Sequences**. As an exercise you may want to extend this chart, predicting all the formulae for say the 12-agonal numbers and onwards and locate where the number (144) Light Harmonic appears.

TYPE	FORMULA	NUMBER SERIES									
(Differences in the Vertical Values)		1	3	6	10	15	21	28	36	45	
TRIANGULAR	$(n^2+n)/2$	1	3	6	10	15	21	28	36	45	55
SQUARE	n^2	**1**	4	9	16	25	36	49	64	81	100
PENTAGONAL	$(3n^2-n)/2$	1	**5**	12	22	35	51	70	92	117	145
HEXAGONAL	$2n^2-n$	1	6	**15**	28	45	66	91	120	153	190
HEPTAGONAL	$(5n^2-3n)/2$	1	7	18	**34**	55	81	112	148	189	235
OCTAGONAL	$3n^2-2n$	1	8	21	40	**65**	96	133	176	225	280
NONAGONAL	$(7n^2-5n)/2$	1	9	24	46	75	**111**	154	204	261	325
DECAGONAL	$4n^2-3n$	1	10	27	52	85	126	**175**	232	297	370
11-AGONAL	$(9n^2-7n)/2$	1	11	30	58	95	141	196	**260**	333	415

Fig 86
THE GEOMETRIC (POLYGONAL) NUMBER SERIES
(BY JAIN 1984 AND R. BROOKS 2001)
highlighting (in bold/underlined) the descending diagonal that is the Magic Square Constants.

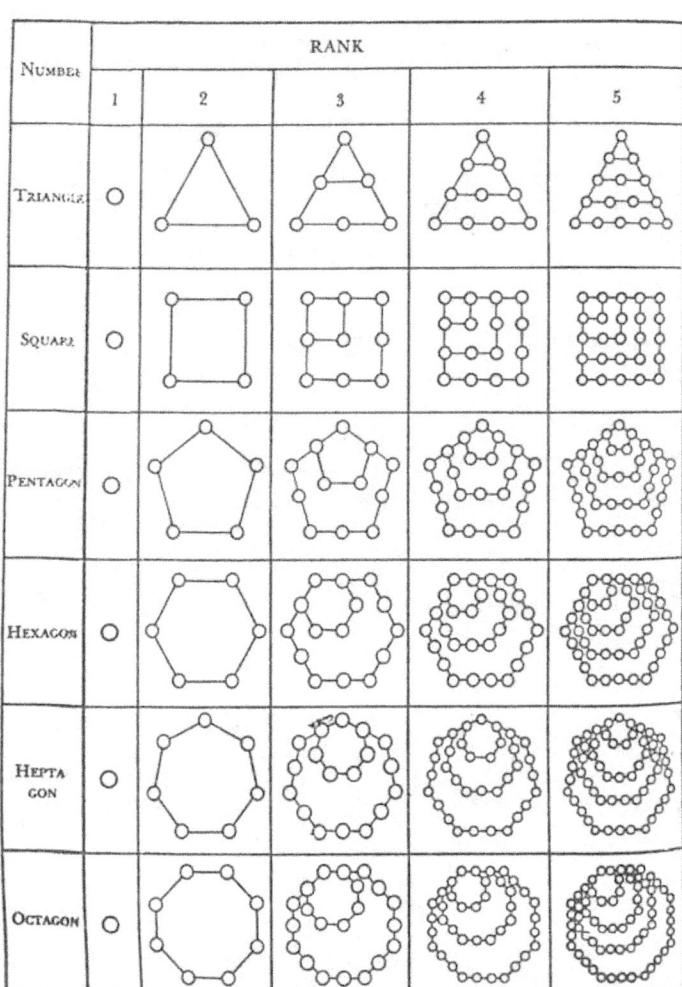

Fig 87
The Polygonal Figurate Numbers.

THE DIMENSIONAL LADDER OF MAGIC SQUARE FORMULAE

To conclude this segment on Magic Squares, as we need to proceed onto the Squared, Triangular, Tetrahedral and Square Pyramid Number Series, I now offer you insight into Time Travel, the ability to move between the worlds in Space and Time.

In a sense, this Magic Square Constants Series, determined by the formula

$$C = [\, n \cdot (1 + n^2)\,] \div 2$$

is very much a vibrational signature for me on Earth or my Space/Time capsule or Vehicle of Light. It is a personal, 2-Dimensional Formula that opens a window into another view or reality or dimension. (In the formula, the mid part of $[1 + n^2]$ represents the First [or 1] and the Last Number $[n^2]$, the <u>Alpha and the Omega</u> principle. Other people will resonate to other formulae, equations, shapes etc as their specific light craft that can transport them inter-dimensionally.

First we have to understand the spectrum of dimensions symbolized by the harmonics of the Magic Square/Cube shapes.

What do you expect the formula would be if we wanted to know the 3-Dimensional Formula for what all the Magic Cubes, not Squares, add up to or vibrate to. Remember that the answer is right infront of you, there is a distinct pattern involved as we begin to climb, Shifting Into The Next Dimension, the Jacob's Ladder of Magic Square / Magic Cube / Magic Multi-Dimensional Constants. The answer for the **Magic Cube Constants** is arrived at by replacing the power or exponent of 2 with a power or exponent of 3, in the above formula. This gives us:

$$C = [\, n.(1 + n^3)\,] \div 2$$

This means, eg: if I wanted to know what the Magic Sum or Constant is for the Magic Cube of 3 x 3 x 3, I substitute the value of n=3 into the formula:
C = [3.(1 + 3³)] ÷ 2
 = **42**.

Have you read the '*Hitchhiker's Guide To The Galaxy*' and the '*Restaurant At The End Of The Universe*' that references the Number 42 as The Secret of The Universe. With the Magic Cube of 3x3x3 having all of its summations in space adding to 42, rows, columns and space diagonals, this places **42** as the first and **primal Harmonic** that allows energies to jump from the 2-Dimensional reality to the 3-Dimensional reality, from the Square to the Cube.

Remember also that the Cube is really 2 inter-digitating or inter-penetrating Tetrahedrons (the 4 corner slices of the cube reveal the Tetrahedron!).

Visualise the Tetrahedron like the Egyptian pyramid at Gizeh. Think of all the vast volumes of knowlege written about it; think of the vast quantity of literature Edgar Cayce wrote about; how much history is embedded in the walls; how much prophesy that is encoded in the symbolism of the ascending and descending passages; think of the stellar connections associated with the layout of the 13 pyramids about the Nile River complex linking directly to the layout of the Orion Constellation; think of the word "**SHAPE**". **SHAPE IS MEMORY**. Crystals have shape and memory. Memory is a compound word, ie: it combines two meanings in the Hebrew language: **MEM** = WATER; **ORI** = LIGHT and you will REMEMBER how Light dances on or travels through water is akin to how the Light of Consciousness flows through the waters of the neuro-transmitters, how the Brain is 90% water (Mem). (Someone once described the Human Being as Water on Two Legs!). This is how Vedic Mathematics flows, like lightning, like "Light Shining Upon The Waters". "Shapes" are a galactic light language of symbols that assist us to think "In The Next Dimension", ultimately to be in All Dimensions by having a universal passport that allows free, connected and unlimited travel.

Understanding how simple these formulae are, can you now predict what the Magic Sum is of the 4th Dimensional Magic Cube of 3 x 3 x 3 x 3 is? It is also known as the HyperCube or Tesseract (see Figs 12 and 58). According to physicist and language of light decoder Dan Winter, another Dimension is just another axis of spin. We don't need to enter Einstein's concept of the 4th Dimension as Time, nor that the Vedic Seers employed Calendars/Time Codes based on Magic Squares which they engraved on precious metals symbolic of the 7 planets and 7 days of the week, but continue on our Dimensional Ladder.

$$C = [\, n \cdot (1 + n^4)\,] \div 2$$

Thus the Magic Sum Constant for the 3x3x3x3 Magic HyperCube is found by putting n=3 into the above formula:

$C = [\, 3 \cdot (1 + 3 \times 3 \times 3 \times 3)\,] \div 2$
$= 123$

The clues are now apparent why you never learnt Magic Squares, Phi and the Platonic Solids and Vedic Mathematics etc at school. To keep you limited. In school books Magic Squares are used to calculate missing cells to determine the Magic Sum, but lack a right-brain/pattern approach. Sadly, if you read the Brittanica encyclopaedia, one edition concluded, after a fascinating account of the History of Magic Squares and how they can be created etc, that they are mere puzzles or trifles and have no educational value or function. Can you believe that such uneducated nonsense gets published globally?

Here are the Dimensions tabulated with their respective Formulae and Multi-Dimensional Constant. The "Constant" also means Timelessness, Eternal, Forever, Fixed Design.

In the following chart, DIM = Dimension. The Constants are given for Magic "Square" of n=3 and n=4 in the Many-Dimensions.

DIM	FORMULA	CONSTANT for	
		n=3	n=4
2nd	$C=n/2 \cdot (1+n^2)$	15	34
3rd	$C=n/2 \cdot (1+n^3)$	42	130
4th	$C=n/2 \cdot (1+n^4)$	123	514
5th	$C=n/2 \cdot (1+n^5)$	366	2050
6th	$C=n/2 \cdot (1+n^6)$	1,095	8194
7th	$C=n/2 \cdot (1+n^7)$	3,282	32,770
8th	$C=n/2 \cdot (1+n^8)$	9,843	131,074
9th	$C=n/2 \cdot (1+n^9)$	29,526	524,290
10th	$C=n/2 \cdot (1+n^{10})$	88,575	2,097,154

Fig 88
THE DIMENSIONAL LADDER
of the Magic Squares with their corresponding
MULTI-DIMENSIONAL CONSTANTS

In my first book: THE BOOK OF MAGIC SQUARES, Volume 1 (1990) I showed how the pattern or yantra for the Magic Square of 3x3 (drawn 4 times upon itself at 45°, as in my Logo) is contained or inherited genetically into the Magic Square of 4x4 (drawn 4 times upon itself at 45°). They are two different and unique patterns but if you look closer, the 3 is nested fractally within the heart of the 4. You can observe this is Fig 89. This means that the essence of the 3 vibration is carried on and contained in the 4 vibration. Why would Nature want this elevation of order or magnitude, from the 3 to the 4? Its about the ability to be **self-similar**, to survive life's changing wavelengths, so that Life is reproduced. A fractal implies a self-similar replication where the size changes, but not the ratio, as in the idealised picture of a tree whose trunk is like the branch is like the twig etc. The ideal **fractal** is the **Phi Ratio**. Its interesting too that when you draw the pattern for the Magic Square of 3x3 at 0° (ie: as it is, with **no** rotation), the 2 large triangles created are visibly in the Phi Ratio, known as <u>Golden Triangles</u>.

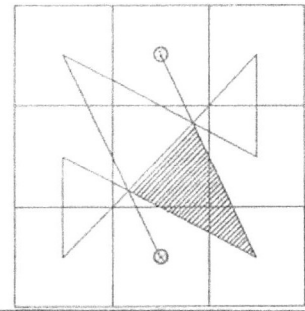

Fig 90
Golden Triangles
of the Lo-Shu.

Fig 89
M.Sq.3 Nested Fractally in the Heart of the M.Sq.4

To conclude this segment, having highlighted the importance of the Recursion of Numbers and their Patterns, how they can be embedded into the heart of greater domains, like the 3 in the 4, we could apply these specific numbers to specific frequencies and electromagnetic devices or Zero Point Technologies that can be used in the arts of healing, relating them to varying spin speeds of the body's chakra system.

Intuitively, the greatest specific Magic Square frequency is the one least obvious and garnishes some of its domain from another Royal Kingdom of Integers known as the Negative Numbers (read as 'minus'). It is possible to arrange a Magic Square using the family of numbers:
–4, –3, –2, –1, 0, +1, +2, +3, +4

Fig 91 shows my favourite Magic Square creation, and, to my knowledge, one that has not been in print before. It is known as JAIN'S ULTIMATE MAGIC SQUARE, for a 3 x 3, whose
- **Constant is Zero**, as the
- 3 vertical columns sum to zero, the
- 3 horizontal rows sum to zero, the
- 2 diagonals sum to zero, the
- 1 central cell is zero, and the
- 4 opposing pairs sum to zero, and the
- 4 corner numbers sum to zero. PERFECTION.

Fig 91
JAIN'S ULTIMATE MAGIC SQUARE: C = ZERO

Part 3a Section 2

the Squared Numbers
(Powers of 2)

referencing also the Cubic Numbers
(Powers of 3)

Section 3 follows this short Section 2 on Squared/Cubic Numbers and conveniently leads us to Triangular Numbers. Let us therefore examine the different ways or expressions of Squared numbers rearranging themselves as Triangular Numbers. If this is possible, would you therefore predict that the Cubic Numbers could rearrange themselves as Tetrahedrons?

Firstly, let us look at the Squared Numbers in their Triangular or Pyramidal form as shown overleaf in Fig 92.

A second way to view The Squared Numbers is to is to arrange them into another pyramidal form (like in Fig 92) but the condition being we list horizontally the consecutive ODD Numbers: 1, 3, 5, 7, 9, etc.

Fig 93 below shows such a pattern that if you really understood its language, your brain could automatically predict the triangular form for 5^2 and recognise "by SHAPE", the specific language of the the Right-Brain Hemisphere, that the Square of 5 is the sum of $1 + 3 + 5 + 7 + 9$ in its triangular form. Or conversely, you could predict that if you were given the task of adding the first 5 odd consecutive numbers, its shape is also a square 5 x 5.

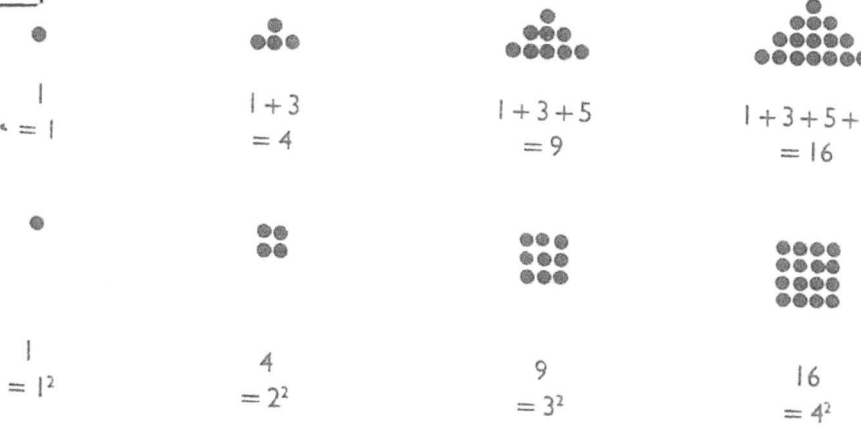

Fig 93
Using SHAPE to Predict the Next Term of Sequences.

Let 'n' represent numbers in arithmetical/consecutive/sequential order ie: 1, 2, 3, 4, etc.
Let 'n^2' be the number squared and expressed below as a **Pyramid of Addition**.
Let 'DS' = the Digital Sum of each row.
Let 'RD' = the Row Difference of each successive number in the n^2 column.

n		n^2	DS	RD
1.	1	1	1	1
2.	1 + 2 + 1	4	4	3
3.	1 + 2 + 3 + 2 + 1	9	9	5
4.	1 + 2 + 3 + 4 + 3 + 2 + 1	16	7	7
5.	1 + 2 + 3 + 4 + 5 + 4 + 3 + 2 + 1	25	7	9
6.	1 + 2 + 3 + 4 + 5 + 6 + 5 + 4 + 3 + 2 + 1	36	9	11
7.	1 + 2 + 3 + 4 + 5 + 6 + 7 + 6 + 5 + 4 + 3 + 2 + 1	49	4	13
8.	1 + 2 + 3 + 4 + 5 + 6 + 7 + 8 + 7 + 6 + 5 + 4 + 3 + 2 + 1	64	1	15
9.	1 + 2 + 3 + 4 + 5 + 6 + 7 + 8 + 9 + 8 + 7 + 6 + 5 + 4 + 3 + 2 + 1	81	9	17

Fig 92
The Squared Numbers Appearing in Their Triangular or Pyramidal Form known as The Pyramid Of Addition

A third way to depict the Odd Numbers as The Squared Numbers is to pack them as "L" formations. If you look closely at Fig 94 below, the SHAPE of the 7 x 7 Square is really the Series of the first 7 ODD Numbers: 1, 3, 5, 7, 9, 11, 13.

In other words, the successive Odd Numbers constitute the differences between each Square Number and the next in the series.

Fig 94
The SQUARE as "L"-SHAPED ODD NUMBERS

A fourth way to show The Squared Numbers is as Triangular Numbers. Specifically, as shown in Fig 95, every Square Number is the sum of two successive Triangular Numbers; the greater Triangular Number corresponds to the order or size of the Square. Here, we have a 7 x 7 Square thus the greater Triangular Number is the Square Root of 49 which is 7, therefore the 7 x 7 Square is composed of the 7th and 6th Triangular Numbers, which is the sum of 28 and 21.

Fig 95
A SQUARE is the sum of two TRIANGULAR Nos.

Buckminster Fuller, whose research on the Platonic Solids in the 1950s and 60s, gave us the geodesic dome, viewed Squares as Rhombuses and Triangular Numbers.

In Fig 96, he shows how Nature prefers the Rhombus Shape, which is like a diamond, in terms of sphere-packing, or closest packed symmetry. (The next diagram of Fig 97 will explore this concept further 'In The Next Dimension', that there is a more efficient system than cubic packing!).

Observe again how B. Fuller highlights the Squared Numbers as Two Triangular Numbers. Can you can imagine all the circles, white and black, as spheres or ping-pong balls that can be glued into triangles and stacked upon each other in appropriate layers. Thus each Squared Number is really two layers of ping-pong balls arranged in triangles. 3 glued balls can nest efficiently in the grooves of the 6 glued balls, giving a total of 9, the Square of 3.

If we extend this concept of triangularly stacking ping-pong balls from 2 layers, to 3 layers, to 4 layers, to 5, 6, 7 etc we will arrive at the Tetrahedral or Tetrahedronal Numbers. This closest packing of spheres is Nature's most **economical** and symmetrical arrangement. It is also the **Path of Least Resistance**.

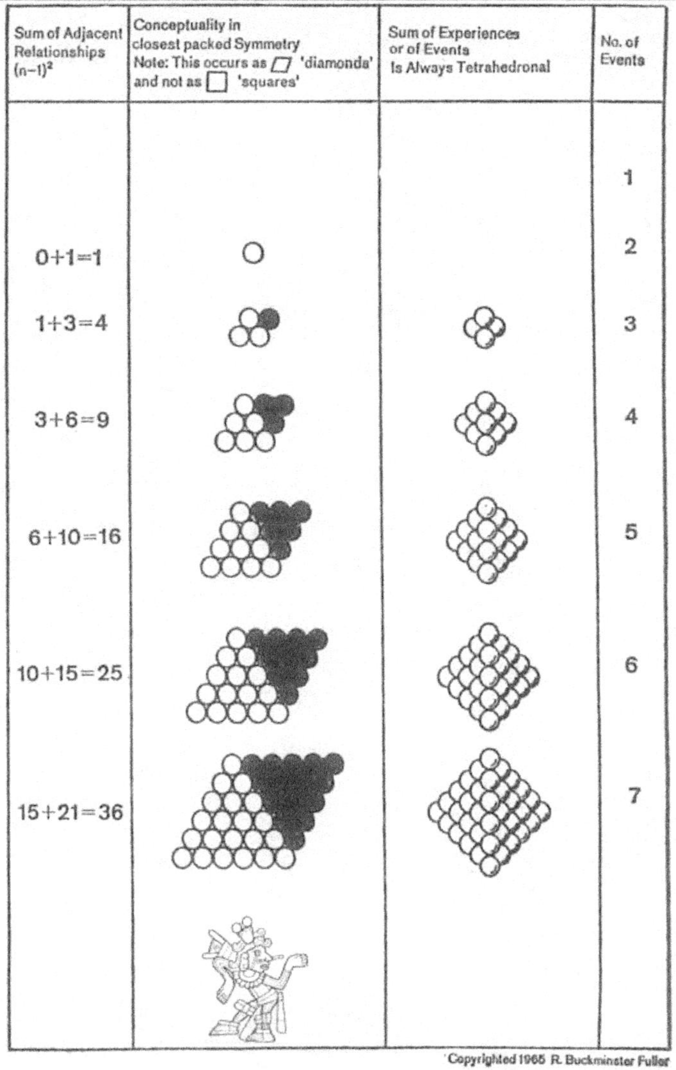

Sum of Adjacent Relationships $(n-1)^2$	Conceptuality in closest packed Symmetry Note: This occurs as ◇ 'diamonds' and not as ☐ 'squares'	Sum of Experiences or of Events Is Always Tetrahedronal	No. of Events
			1
$0+1=1$	o		2
$1+3=4$	◦◦◦◦	◦◦◦◦	3
$3+6=9$	◦◦◦◦◦◦◦◦◦	◦◦◦◦◦◦◦◦◦	4
$6+10=16$			5
$10+15=25$			6
$15+21=36$			7

Copyrighted 1965 R. Buckminster Fuller

Fig 96
Squares as Closest-Packed Symmetrical Systems

Buckminster Fuller also argued that our habitual references to squares and cubes have kept us locked into a right-angled viewpoint, thus obscuring our vision of truth.

"From now on," says Bucky, "we have to say 'triangling' not 'squaring' if we want to play the game the way nature plays it".

What he means by this is that the cube is not superior to the tetrahedron. If you sit on a cubic box, the forces are weak enough that it will distort into a rhombus and collapse. In fact, the Cube is very weak in contrast to the Tetrahedron which is the basic and **most stable** building block of the Universe.

(The tetra has 4 faces and **4 vertices**. Two Tetras interpenetrate to form a cube with 8 or 2x4 vertices. 3 Tetras interpenetrate to form the icosahedron with 12 or 3x4 vertices. 5 Tetras interpenetrate to form the dodecahedron with 20 or 5x4 vertices)

Fig 97 shows graphically this concept of "Cubing" versus "Tetrahedroning".

The diagram shows the obvious 1, 8, 27 and 64 Cubelets that make up the larger Cubes, but if you look closer, the corresponding Tetrahedrons have the same number of smaller tetrahedrons as do the Cubes.

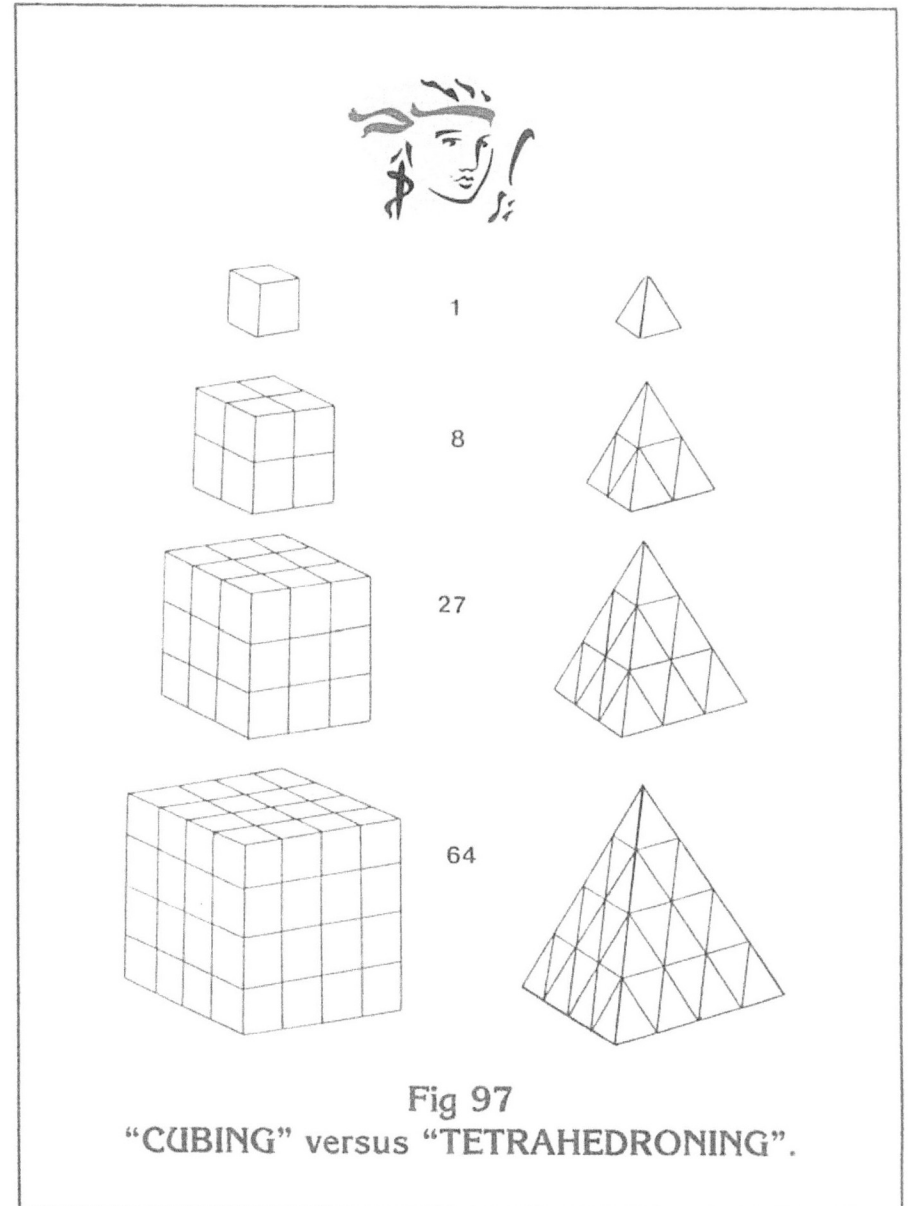

Fig 97
"CUBING" versus "TETRAHEDRONING".

Part 3A Section 3
The Triangular Numbers

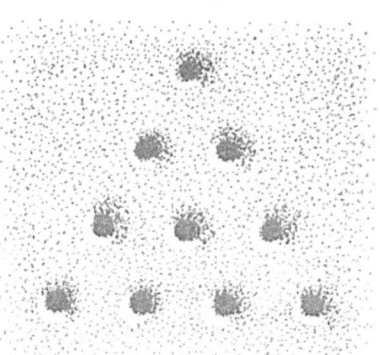

Fig 98
Pythagoras' TETRAKTYS
"NUMBER IS THE BOND OF ALL THINGS"

Perhaps the simplest definition of the Triangular Series of numbers is the Vedic Sutra: "By One More". The simplest arithmetic series of numbers is just to keep adding the next "natural" number:

1
1+2
1+2+3
1+2+3+4
1+2+3+4+5 etc

If you wanted to know the next Triangular Number, just add 6 to the last sum. What if you wanted to know the 20th Triangular Number? <u>Formulae are short-cuts that recognise Universal Patterns</u>. The Formula needed here is:

n(n + 1) / 1x2

This means that to find the 20th term, let n=20. We multiply this 20 by the next natural number which is 21 and divided this amount by 2.
=20 x 21 / 2
=210

In The Next Dimension, The Tetrahedral Numbers, to be discussed next, obey a similar pattern:

n(n + 1).(n + 2) / 1.2.3

(The "Period" or "dot" above refers to multiplication).

Likewise, by understanding the patterns in the formulae in the 2nd and 3rd Dimensions, you could easily predict what we will call the 4th Dimensional (Δ) Triangular Series: 1, 5, 15, 35, 70 etc conforming to this shape:

$$n.(n + 1).(n + 2).(n + 3) / 1.2.3.4$$

Euclid (fl.c.295BC, Greek) ("fl" = flourished, and "c" = circa or about) wrote many volumes on mathematics which have become part of our curriculum, and it is full of the world's inherited knowledge on theorems and postulates. In it are typical statements like:

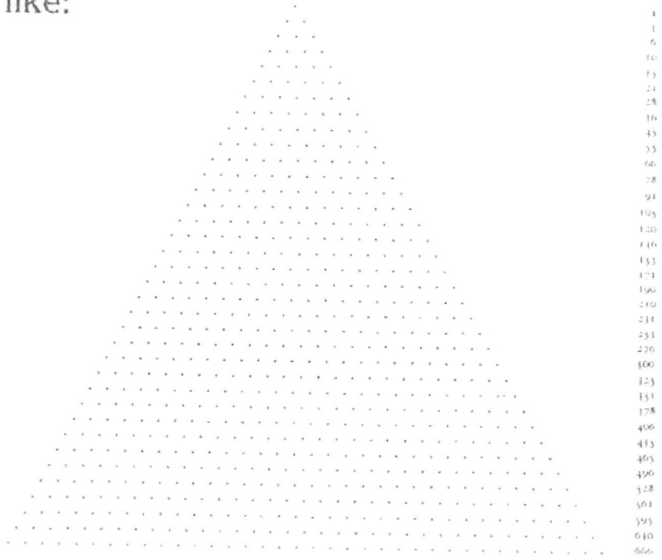

Fig 99
The 36th Triangular Number = 666

"THE SUM OF TWO CONSECUTIVE TRIANGULAR NUMBERS IS A SQUARE NUMBER".

Pythagoras (c.580 - c.480BC, Greek) surely knew this but but modern text books would attribute this discovery to the 4th century by the Greek scholar Theon, father of the first known woman mathematician Hypatia.

Fig 100 allows the student to instantly predict the next number in the series:

$1 + 3 = 2 \times 2$
$3 + 6 = 3 \times 3$
$6 + 10 = 4 \times 4$
$10 + 15 = ?$

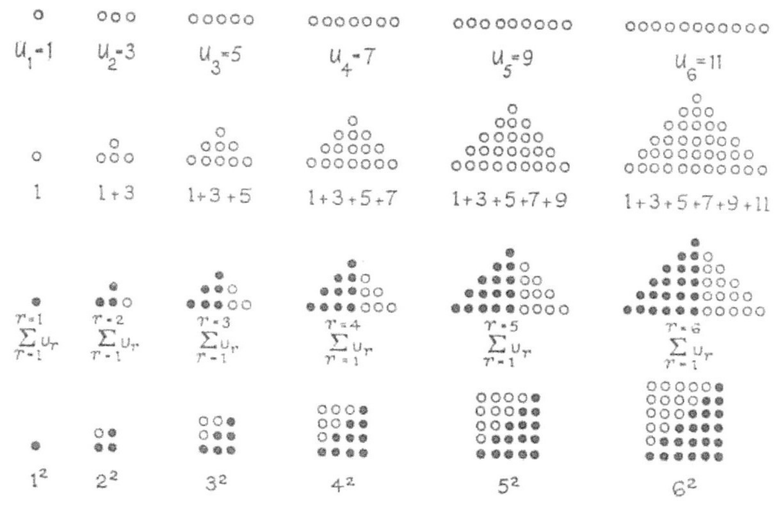

Fig 100
Two Consecutive Triangular Nos. Make a Squared No.

HOW GAUSS SOLVED THE FORMULA FOR THE TRIANGULAR NUMBERS.

There is a story told of a great German Mathematician Karl Gauss, who solved a problem at the age of 10, in a very smart way (see Fig 101 below). When his teacher asked him to add $1 + 2 + 3 + 4 \ldots + 18 + 19 + 20$ Gauss quickly gave the answer. He had 3 lines of setting out:

1. $1 + 2 + 3 + 4 + 5 + \ldots + 16 + 17 + 18 + 19 + 20$
2. $20 + 19 + 18 + 17 + 16 + \ldots + 5 + 4 + 3 + 2 + 1$
---
3. $21 + 21 + 21 + 21 + 21 + \ldots + 21 + 21 + 21 + 21 + 21$

Fig 101
Gauss' Method of Adding Numbers

Gauss wrote the second line the same as the first line but in reverse order. This makes apparent that the third line is 20 pairs of 21 or 20 x 21. Thus, the desired sum is half of this or (20 x 21)/2 = 210. (Remember, this is the formula for the Triangular Numbers).

This account is merely to demonstrate that when we don't know a specific formula, there are many ways of approaching the problem, not always in the way we expect to solve an operational problem, but most successfully when we dissect the data with the keen X-Ray eyes of a Pattern Hunter.

To conclude this section on Triangular umbers, use Shape and Pattern, in a Gaussian sense, to instantly predict the missing number in the following series, Fig 102. Then write out your own formula from this observation.

$1^3 = 1 = 1^2$
$1^3 + 2^3 = 1 + 8 = 3^2$
$1^3 + 2^3 + 3^3 = 1 + 8 + 27 = 6^2$
$1^3 + 2^3 + 3^3 + 4^3 = 1 + 8 + 27 + 64 = 10^2$
$1^3 + 2^3 + 3^3 + 4^3 + 5^3 = 1 + 8 + 27 + 64 + 125 = ?$

Fig 102
Sums of the Consecutive Cubic Numbers.
Predicting the Missing Number
and Defining the Discovery of a Universal Pattern.

One of the 16 Vedic Sutras is knowing an answer "<u>By Mere Observation</u>", "<u>By Mere Intuition</u>".

This book is the 17th Sutra, the Lost Sutra, that steps awarely into the Next Dimension, to activate the 4th Dimensional Eye or All-Knowing Eye. What I am teaching here, is that we already know the answer, like the Rain Man, as soon as it is given. THE SUM OF CONSECUTIVE CUBES = THE SQUARE OF A TRIANGULAR NUMBER. The answer must therefore be **15 Squared**, 15 being the 5th Δ No.

Part 3A — Section 4

THE TETRAHEDRAL OR TETRAHEDRONAL NUMBERS :

1, 4, 10, 20, 35, 56, 84, 120, 165, 220, 286

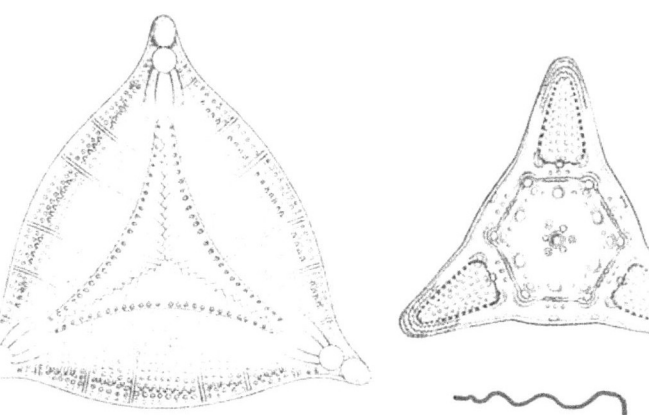

Tetrahedral-shaped Diatom: type of unicellular plant.

The following whole page is reproduced from my book: THE MAGIC OF NINE IN VEDIC MATHEMATICS, 2001, from page 111. It was expected to be published in 2002 but in that year I wrote: THE BOOK OF PHI, Volume 1, 2002 as the precursor to this publication.

Having come from a Magic Square background, I have known for several decades the revelation (Fig 103) that the Magic Sums of the Magic Squares: (**15, 34, 65, 111, 175, 260, 369**) are encoded triplically as the Tetrahedral Numbers:

- 1 + 4 + 10 = **15**
  = the Magic Square Constant of 3 x 3.
- 4 + 10 + 20 = **34**
  = the Magic Square Constant of 4 x 4.
- 10 + 20 + 35 = **65**
  = the Magic Square Constant of 5 x 5.
- 20 + 35 + 56 = **111**
  = the Magic Square Constant of 6 x 6.
- 35 + 56 + 84 = **175**
  = the Magic Square Constant of 7 x 7.
- 56 + 84 + 120 = **260**
  = the Magic Square Constant of 8 x 8.

Fig 103
3 Consecutive Tetrahedral Nos. = M. Sq. Constants.

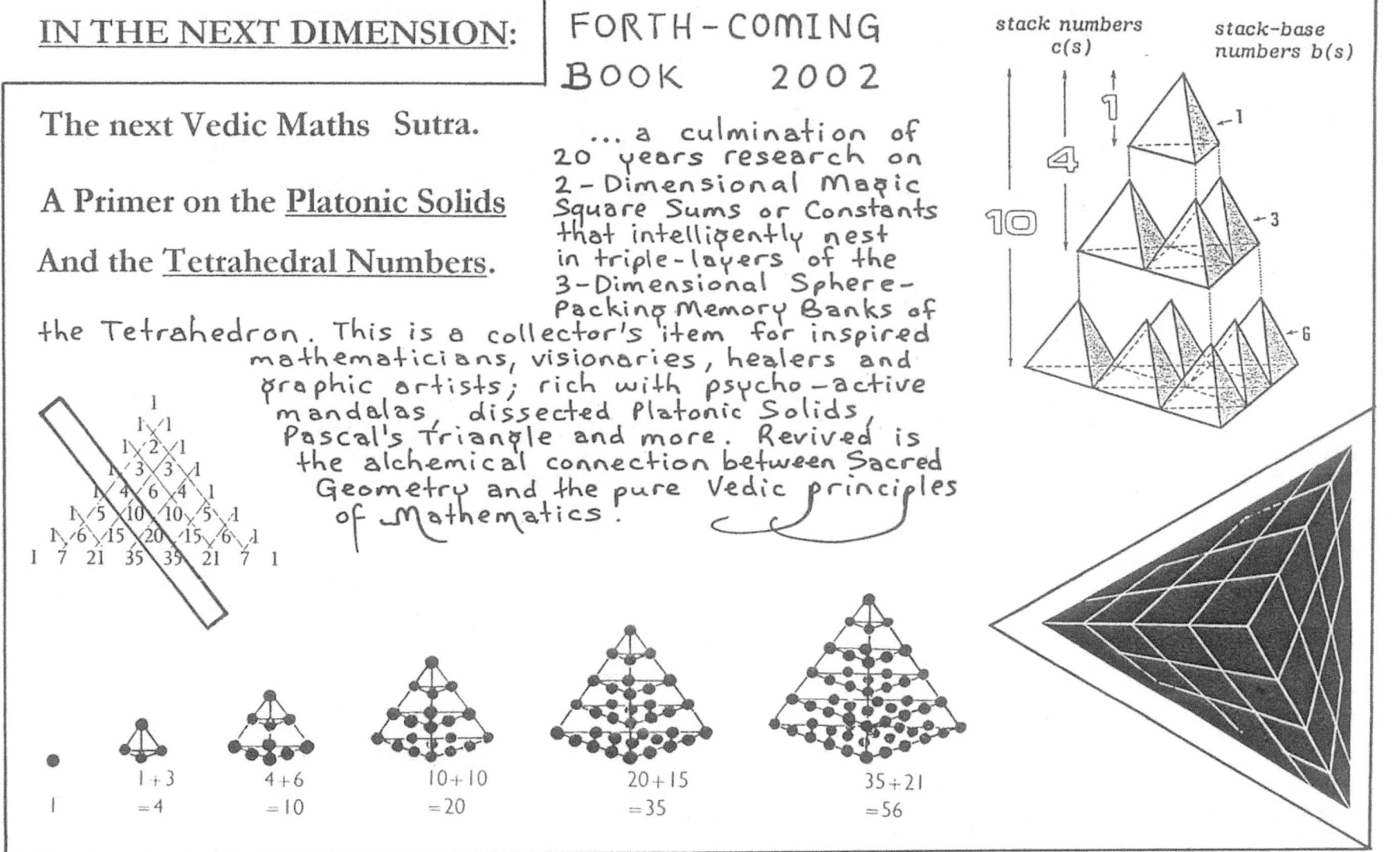

IN THE NEXT DIMENSION:

The next Vedic Maths Sutra.

A Primer on the **Platonic Solids**

And the **Tetrahedral Numbers**.

FORTH-COMING BOOK 2002

... a culmination of 20 years research on 2-Dimensional Magic Square Sums or Constants that intelligently nest in triple-layers of the 3-Dimensional Sphere-Packing Memory Banks of the Tetrahedron. This is a collector's item for inspired mathematicians, visionaries, healers and graphic artists; rich with psycho-active mandalas, dissected Platonic Solids, Pascal's Triangle and more. Revived is the alchemical connection between Sacred Geometry and the pure Vedic principles of Mathematics.

EXCERPT from "THE BOOK OF PHI" Volume 1, that announced the Tetrahedral connection to Magic Squares. 2002

The method by which the Tetrahedral Numbers are generated is by consecutively stacking the Triangular Number Series:

1, 3, 6, 10, 15, 21, 28, 36, 45, 55 etc

into an ever-increasing tetrahedron. This is shown below in Figs 104 and 105. Imagine again that all the little circles in the diagrams are ping-pong balls glued or connected.

The meaning of consecutive means we add from 1, successively each time, in order of their appearance. Thus:

```
1                        = 1
1 + 3 =                  = 4
1 + 3 + 6 =              = 10
1 + 3 + 6 + 10 =         = 20
1 + 3 + 6 + 10 + 20 =    = 35 etc
```

$T_{r,3} = \dfrac{r(r+1)(r+2)}{6}$

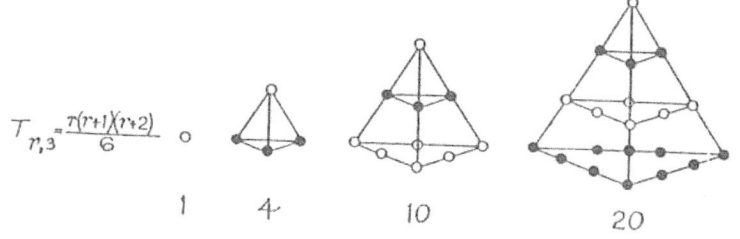

Fig 104
The First 4 Tetrahedral Numbers.

Our Formula is:
<u>THE SUM OF CONSECUTIVE TRIANGULAR NUMBERS IS ALWAYS A TETRAHEDRAL NUMBER.</u>

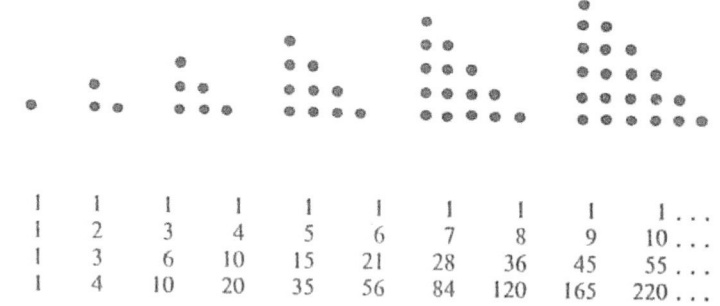

1	1	1	1	1	1	1	1	1	1 ...
1	2	3	4	5	6	7	8	9	10 ...
1	3	6	10	15	21	28	36	45	55 ...
1	4	10	20	35	56	84	120	165	220 ...

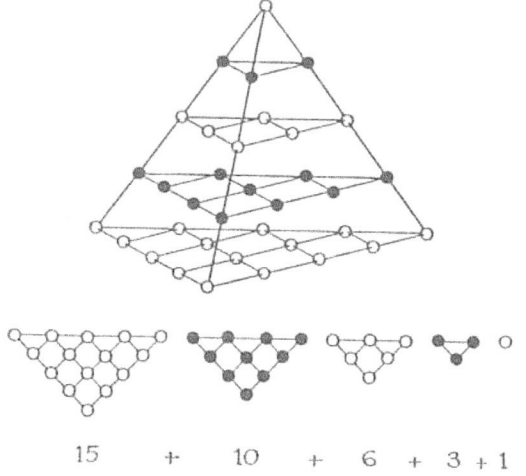

15 + 10 + 6 + 3 + 1

Fig 105
Analysis of the 5th Tetrahedral Number = 35.

I offer here a symbolic picture that aptly captures the idea of the 3-Dimensional Series of Tetrahedral Numbers which is born out of the 2-Dimensional Tri∆ngular Series.

This is shown in Fig 106 below, on the right-hand-side, generating parabolic curves artistically using only straight lines, a process called "Curve Stitching" which you may have experimented with at school.

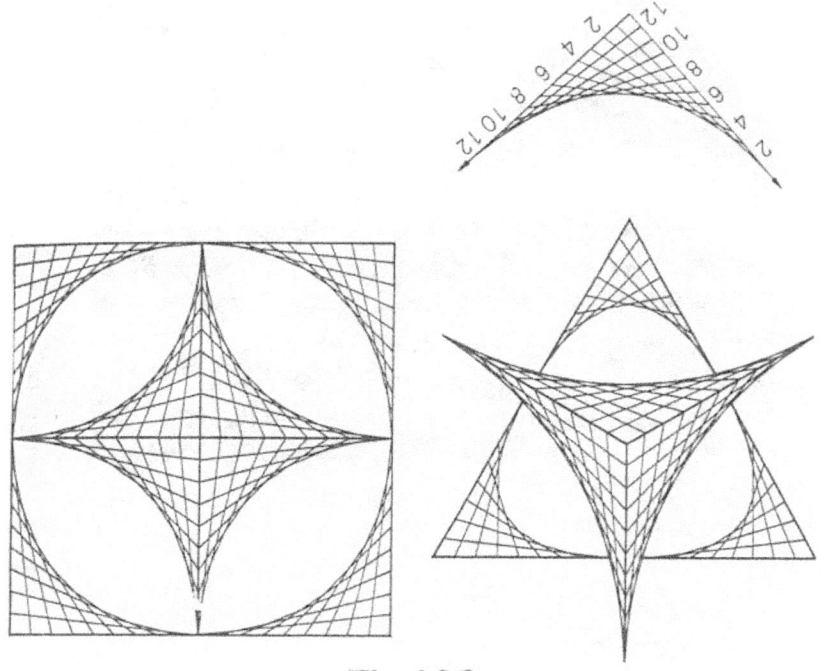

Fig 106
"Curve Stitching" Tetra born out of Triangle.

If you wanted to know the 6th Tetrahedral Number, substitute n=6 into the Formula:

$$n.(n + 1).(n + 2) / 1.2.3$$

= 6x(6+1)x(6+2) ÷ 1x2x3
= 6x7x8 ÷ 6
= 56

Another way of viewing these stacked numbers is shown below in Fig 107. Buckminster Fuller referred to this diagram as the TETRAHEDRON STACK NUMBERS which is really referring to the Tetrahedral Series. Some old maths books refer to Fig 107 as a SOLID TESSELLATION.

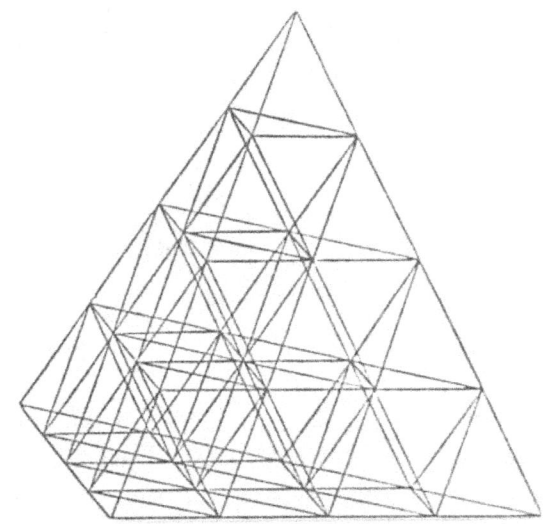

Fig 107
THE TETRAHEDRON STACK NUMBERS. B.Fuller.

The world's smallest satellite has been developed by Space Technology Laboratories. Its shape will be different from all other satellites before it. STL engineers and scientists have used a <u>tetrahedral configuration</u> to bring about some remarkable characteristics in a space vehicle. There will be no need for batteries or regulators in flight. The satellite will have no hot side, no cold side. It will require no altitude control devices. No matter how it tumbles in space it will always turn one side towards the sun to absorb energy, and three sides away from the sun to cool instrumentation and telemetry equipment inside.

Fig 108
The TETRAHEDRON Design of a Space Satellite.

The following diagrams (Figs 109, and 110) are taken from Chris Illert's book: <u>ALCHEMY TODAY</u> "Platonic Geometries in Nuclear Physics", Volume 1, pub. 1992, which commemorates Leadbeater's and Besant's classic: <u>OCCULT CHEMISTRY</u>.

Fig 109 shows the number of alpha-particles, c(s), in a compressed core of nuclear matter belonging to a nucleus with 's' main shells. **These core numbers correspond to Buckminster Fuller's tetrahedron stack numbers.**

Basically, Illert's and Leadbeater's physic's contradicts conventional 'quantum physics' in that they recognise the composition of matter as a nesting of the Platonic Solid symmetries. eg Calcium (40) may be viewed as a central Tetrahedron (T) within an Icosahedral (I) Oxygen nucleus and both these are nested within a special type of Dodecahedral outer shell (DT) having 24 distinct vertices or nucleons. Its configuration or code is therefore T, I, DT.

s	Buckey Fuller's tetrahedron stack numbers c(s)
1	$0^2 = 0$
2	$1^2 = 1$
3	$0^2 + 2^2 = 4$
4	$1^2 + 3^2 = 10$
5	$0^2 + 2^2 + 4^2 = 20$
6	$1^2 + 3^2 + 5^2 = 35$
7	$0^2 + 2^2 + 4^2 + 6^2 = 56$
8	$1^2 + 3^2 + 5^2 + 7^2 = 84$
9	$56 + 8^2 = 120$
10	$84 + 9^2 = 165$

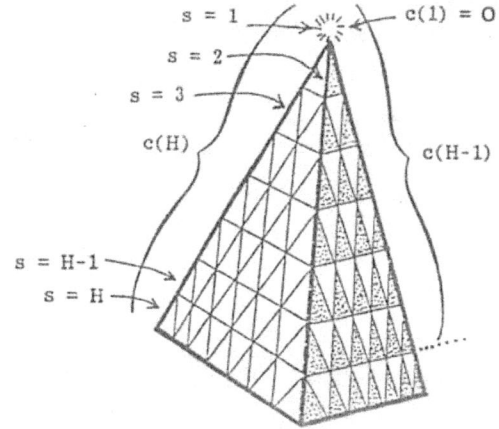

Fig 109
The Tetrahedral Numbers
Appearing in Leadbeater's Nuclear Geometry.

Fig 109 shows equal-sized tetrahedra stacked on top of each other, in piles of arbitrary height. The cumulative total number of tetrahedra in any such pile, is called the TETRAHEDRON-STACK NUMBER and it can assume the values of: 0, 1, 4, 10, 20, 35, 56... etc (these are The TETRAHEDRAL NUMBERS) depending upon how many layers (s-1) are in the stack. We therefore denote it as "c(s)" to show that it depends on the variable or parameter "s". The corresponding number of tetrahedra in the base-layer alone, is called the STACK-BASE NUMBER and is denoted by "b(s)" because it too depends upon "s" (the height of the stack). Stack-base numbers may assume the values of: 1, 3, 6, 10, 15, 21, 28... etc (which are of course The TRIANGULAR NUMBERS).

It was Buckminster Fuller who initially proposed using tetrahedron stacks to represent shells of nucleons in dense-packed nuclear cores. He never knew just how right he was!

Fig 110 shows detailed models of the structure of various "magic" atomic nuclei, in terms of the nested Platonic Shells, according to Buckminster Fuller in the 1930's to 1950's...before quantum physicists had discovered how to generate realistic shells through introduction of a "spin-orbit" coupling term in their Schödinger "wave-equations".

All nuclear cores can be represented as tetrahedron stacks. In the 1930's, Fuller didn't know the subshell finestructure in compressed nuclear matter, so he simply represented the nuclear core by a stack of tetrahedra (alpha-particles or "<u>ADYARIA</u>") surrounded by various "valence" Platonic subshells. See also Figs 42-48 for more.

After R. Buckminster Fuller's book "synergetics"

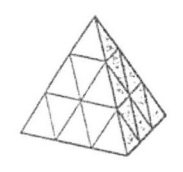
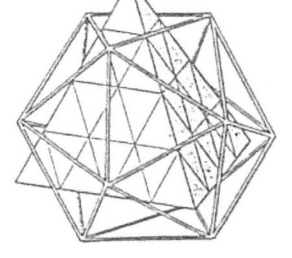

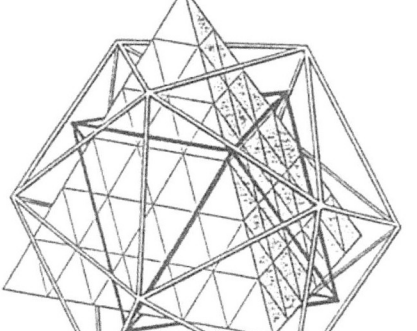
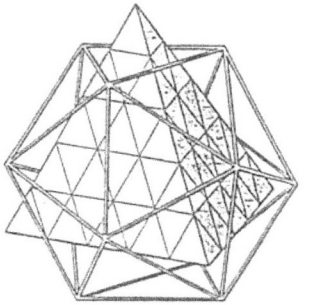

Fig 110
Nuclear Cores represented as Tetrahedron Stacks.

The following whole page shows the setting out for the original revelation regarding these Tetrahedral Numbers and how they combine in triadic sums to generate the Magic Sums of the Magic Squares. This is shown as Fig 111. (nb: by "co-incidence" 111 is also the Magic Square Constant for the 6 x 6 Magic Square of the Sun). Research the work of **Prof J.R. Searl**, "<u>THE LAW OF THE SQUARES</u>", who used these Magic Squares to develop "free energy" systems like space-craft with vertical take-off. (For more info on Prof. Searl, see my book: THE BOOK OF MAGIC SQUARES, Vol. 2, pub. 2000, intro).

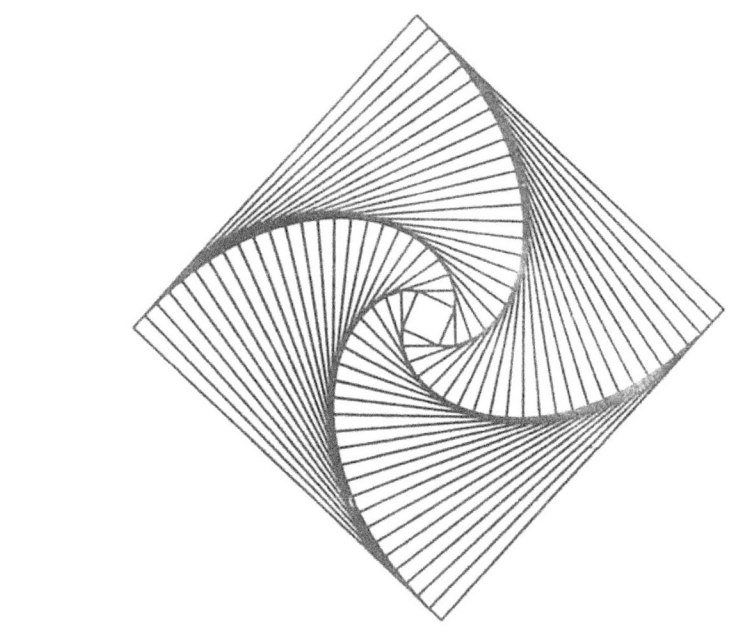

n	Formula	TN	Magic Square Constants as the Sum of 3 Tetrahedral Numbers
1.	1.2.3 / 6	= 1	
2.	2.3.4 / 6	= 4	= 15
3.	3.4.5 / 6	= 10	= 34
4.	4.5.6 / 6	= 20	= 65
5.	5.6.7 / 6	= 35	= 111
6.	6.7.8 / 6	= 56	= 175
7.	7.8.9 / 6	= 84	= 260
8.	8.9.10 / 6	= 120	= 369
9.	9.10.11 / 6	= 165	= 505
10.	10.11.12 / 6	= 220	= 671
11.	11.12.13 / 6	= 286	

Fig 111
Jain's Original Revelation, in 1983, that the Magic Sums of the Magic Squares are Consecutive Triplets or Triadic Sums of the Tetrahedral Numbers (T N) Series.

Many 'Planetary Pioneers', whether they are scientific or mystical, are devising ways of **restructuring water** so that it reclaims its original Platonic Solid Symmetry Memory. **Professor Flanagan**, who invented the neurophone to assist deaf people to hear again, has a product called Flanagan Microclusters which create **dodecahedral** liquid crystal structures in drinking water. When water is structured with "Crystal Energy", the water molecules form a matrix based on the Dodecahedron (Fig 112 below). This liquid crystal matrix has the unique ability to capture and hold negatively charged Hydrogen Ions, thus delivering significant amounts of electrical energy to the body.

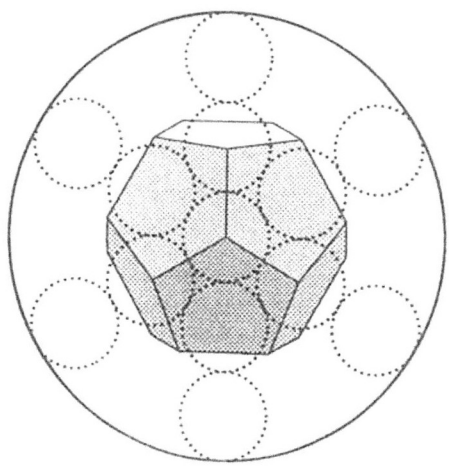

Fig 112
Restructured Water with Dodecahedral Memory.

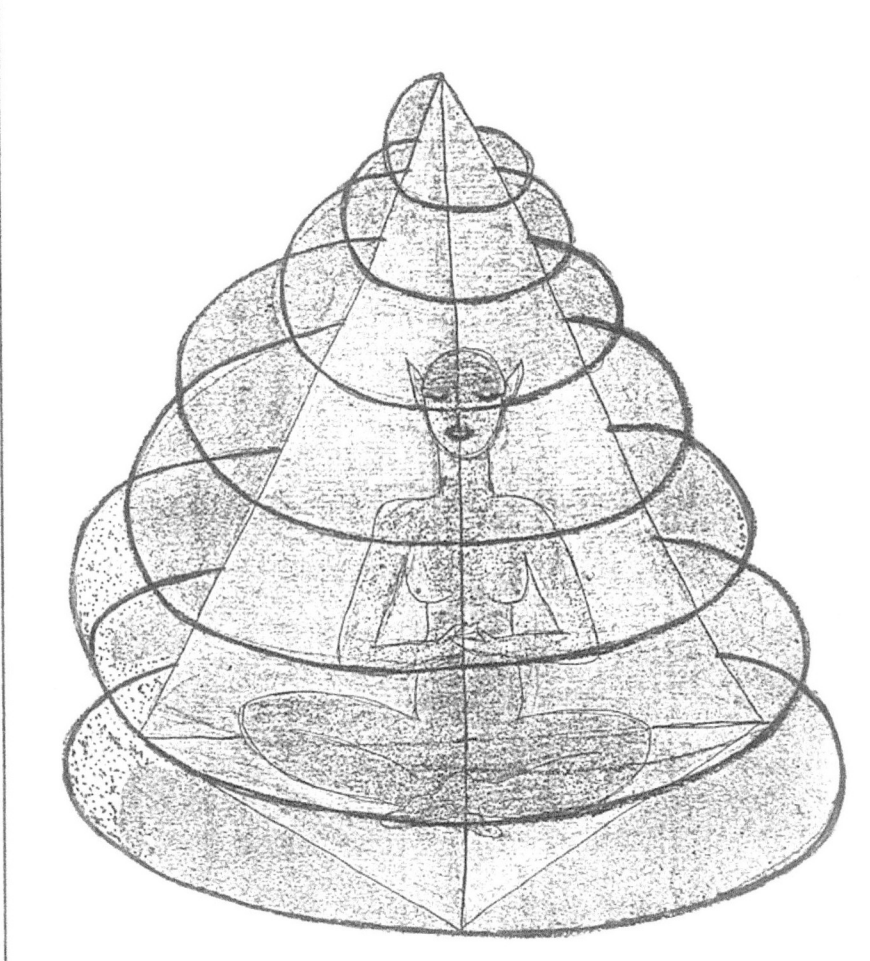

Fig 113
Continue to meditate inside these Platonic Solid Geometries, and a Door will Open.
"TETRAHEDRAL CONSCIOUSNESS" by Jain, 1997.

THE MAGIC TETRAHEDRON

Whilst on the subject of Tetrahedral Numbers, let us look at 3 more interesting perspectives of the Tetrahedron:
1: The Magic Tetrahedron
2: The Tetrahedral Numbers in Pascal's Triangle.
3: Tetrahedral Micro-Fossils of Spores and Pollen.

There are many books with unusual forms of the magic squares and some are given for the reader to fill in the missing numbers.(as in Fig 114). Such is the case with the MAGIC TETRAHEDRON.

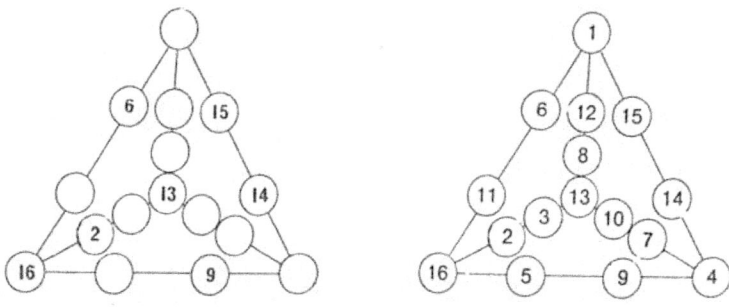

Puzzle

Fig 114
Fill In The Missing Numbers of the Magic Tetrahedron.
+ Solution to the MAGIC TETRAHEDRON

What is of interest here is that the solution is really the 2-Dimensional Magic Square of 4 x4 which knows how to turn itself "inside-out" and collapse efficiently into the 3-Dimensional Tetrahedron such that its <u>Properties include</u>:
- the sum of the 6 edges = 34.
  eg: 1 + 6 + 11 + 16
- the 4 vertices = 34
  eg: 1 + 13 + 4 + 16
- the sum of the 4 corner segments = 34.
  eg: 1 + 6 + 12 + 15
- the 3 central edge Nos. + their complementary central base Nos. = 34
  eg: (12 + 8) + (5 + 9) = 34
- There are 14 summations of 34.
- (There are 4 hexagonal faces of differing values).
- Since there are 880 permutations or different Magic Squares of 4 x 4, then there are also 880 such Magic Tetrahedrons!

See also Figs 58 + 59 + 85:
1 of 880 2-Dimensional Magic Squares of 4 x 4.

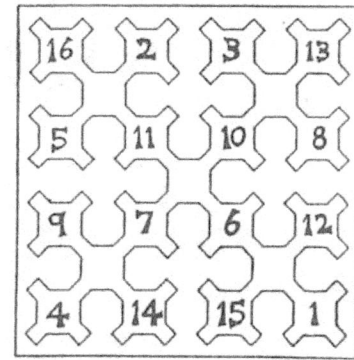

PASCAL'S / HALAYUDHA'S TRIANGLE

The final segment of the Tetrahedral Numbers is the theme of rising to the next dimension, from 2-D to 3-D. Similar to how a 2-D Magic Square is an encoded/encrypted Magic Tetrahedron, there exists another 2-D Pascal's/Halayudha's Triangle that contains not only the 3-D Tetrahedral Numbers but also the universe of Triangle Numbers, Fibonacci Numbers, all the Powers of 2 and all the Powers of 11, and more!

Also known as the Chinese Triangle, any number, like 6 is formed from the addition of two numbers directly above it, which is 3 + 3, as seen in Fig 115.

Notice in Fig 115 the fourth diagonally sloping series of numbers and it is the infamous Tetrahedral Numbers.

For more info, please refer to Figs 34 and 74.

THE HOCKEY STICK RULE:

If you wanted to know the sum of say the first 4 Tetrahedral numbers, which is 1 + 4 + 10 + 20, the answer is given in the 5th diagonally sloping line, in a "L" shape or hockey-stick shape, Fig 116. See where the number 20 is, go down, and observe the number 35 which is the required sum or answer.

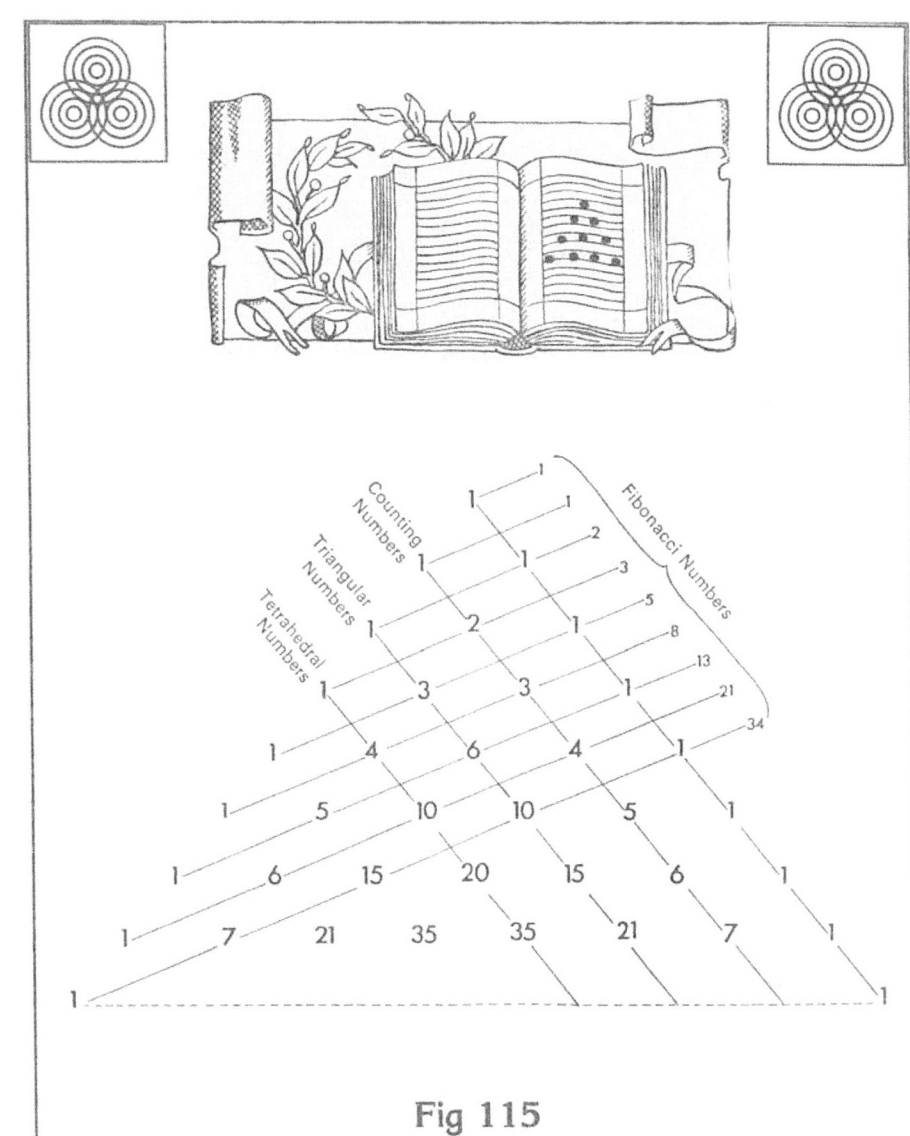

Fig 115
The Tetrahedral Nos. in Pascal's/Halayudha's △.

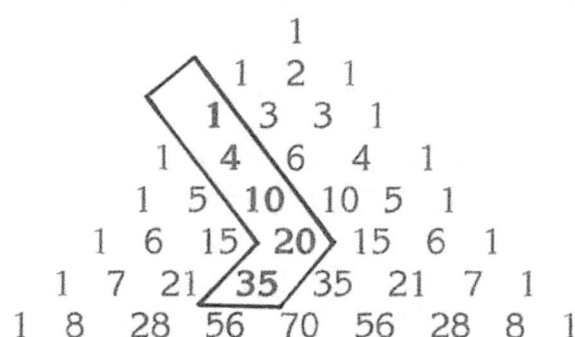

Fig 116
Hockey-Stick Rule to Add Up Tetrahedral Nos.

There is a marvellous 3-D version of this same triangle known as <u>Halayudha's Tetrahedron</u> or <u>Pascal's Pyramid</u>, where each face is a Pascal's Triangle, and all interior numbers are the sum of the 3 numbers directly above it. See Fig 117. In the higher scheme of mathematics, each level of numbers in this pyramid relate to the coefficients in the expansion of a trinomial expressed as (a + b + c) to the 'nth' power.

I reference it here, as it dovetails into this fascination and evolution of the 2-D Triangle becoming the 3-D Tetrahedron, a crossing of the dimensions. If you can walk this bridge, and understand your steps, you can tap into the 4th Dimensional Pascal's / Halayudha's Triangle.

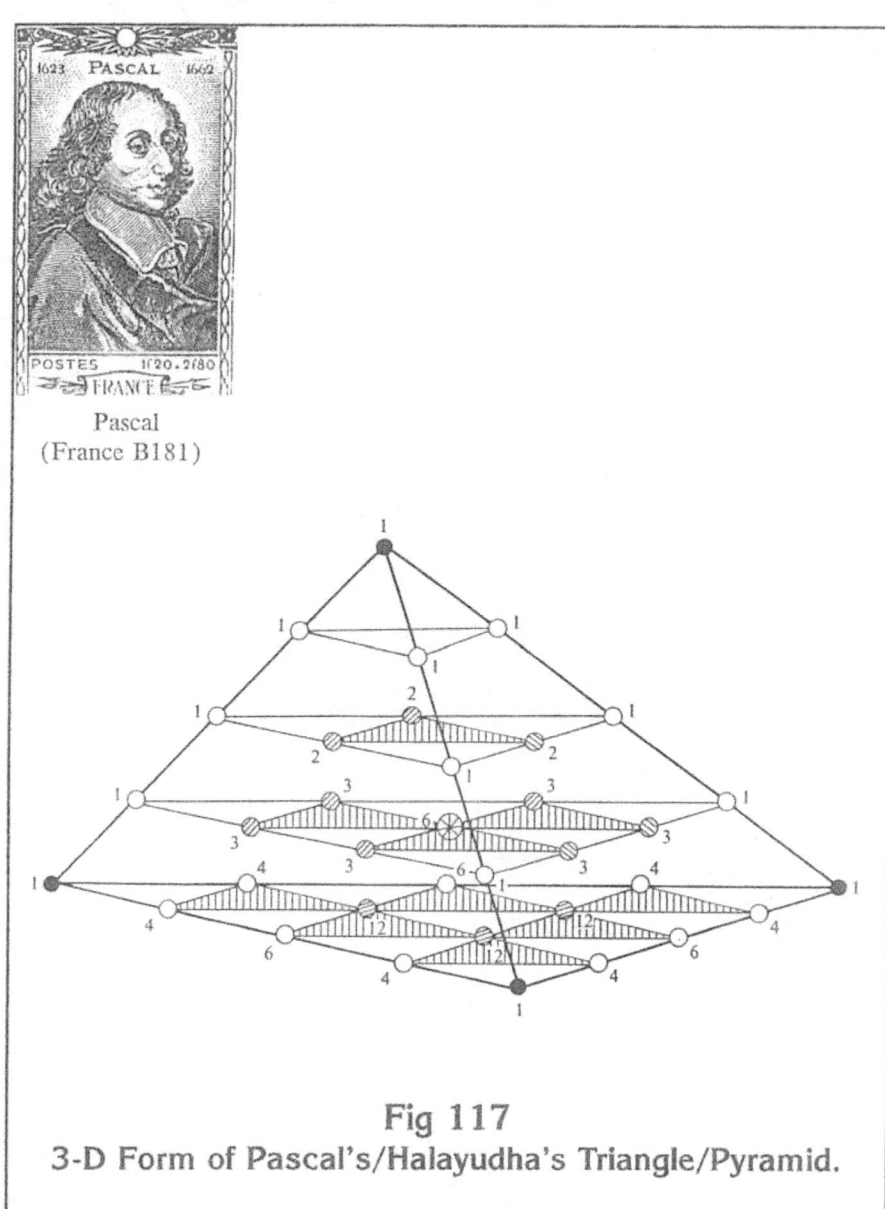

Pascal
(France B181)

Fig 117
3-D Form of Pascal's/Halayudha's Triangle/Pyramid.

Tetrahedral Micro-Fossils of Spores and Pollens

Spores are the reproductive bodies of the primitive non-flowering plants, whereas pollens are the male germinant bodies of the authentic flowering plants. They occur in marine and non-marine bog sediments as far back as the Jurassic Period. Fig 118 shows micro-fossilised Paleozoic Spores magnified from X200 to X450, drawn diagrammatically to show their "<u>trilete sutures</u>" which gives them their distinct tetrahedral morphology. These sutures or triangular lines of junction facilitate the male reproductive cells to escape during the process of fertilisation. Pollens differ in that they are more furrowed and porous as shown below in Fig 119.

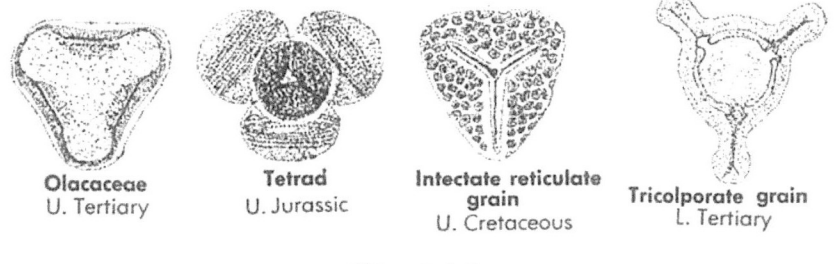

Fig 119
Tetrahedral MicroFossilised Pollen

Fig 118
Tetrahedral Micro-Fossilised Spores

THE ORACLE OF DELPHI

Kings and Queens from afar came to consult with the most famous of all "priestess channels" who was perched on the summit of a <u>TETRAHEDRAL Tripod</u> above the continuous rising of fumes, where the Spirit dwelled, from a gaping crevasse in an ancient Greek cave that appeared to put her in a trance state. Loud clangings could be heard, and the tripod vibrated until the daemon or spirit had taken control. In this altered state, a King may have asked her of the outcome of a certain war or affair, etc and she could see all time, In The Next Dimension.

The three legs of the clawed tripod represent the three periods of Time ruled by Lord Apollo, the Sun God, namely the Past, the Present and the Future. The space defined by the legs of the clawed tripod is the tetrahedron most sacred to the Pythagoreans. The prophetess being enthroned at its apex, above the abyss, is symbolic of our spiritual nature to dominate over the lower worlds.

The face of Apollo is visible and dominates over the circularly coiled python at the base indicating the transmutation of sexual desires towards a higher tantric goal.

Part 3a — Section 5

The Square Pyramid Numbers:

1, 5, 14, 30, 55, 91, 140, 204

+

A 3rd And Original Method Of Realising The Magic Square Constants:

15, 34, 65, 111, 175, 260, 369

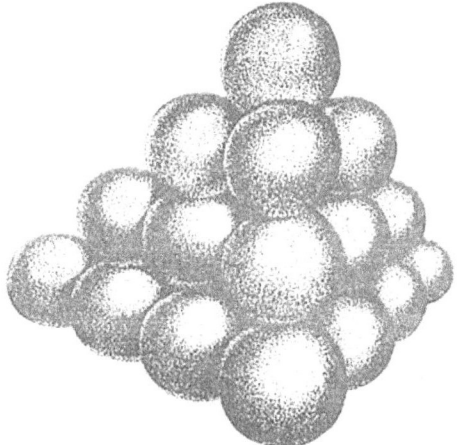

We saw in Fig 94 how the squared numbers were formed by adding successive odd numbers in "L-Shapes". It so happens that when you add consecutive Squared Numbers, beginning from 1, the results are always the PYRAMIDAL NUMBERS, in which each successive layer is a squared number, thus, the 6th Pyramidal Number is:

$1^2 + 2^2 + 3^2 + 4^2 + 5^2 + 6^2 = 91$.

This is shown below as Fig 121. Notice that each face is a Triangular Number.

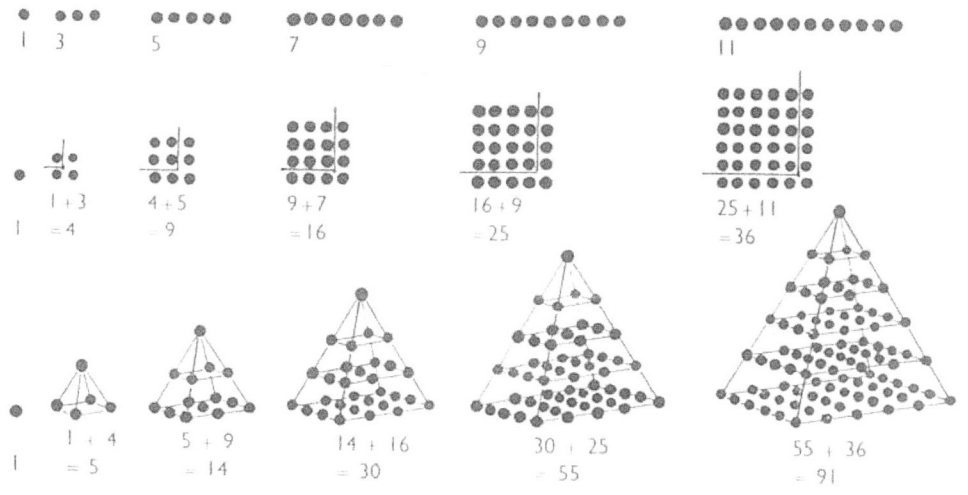

Fig 121
Formation of the 6th Pyramidal FIGURATE Number

Figurate numbers are derived from distinct geometrical shapes as in the squared, triangular, tetrahedral, pyramidal and many other polygonal numbers.

A particular style of stacking spheres is known as the Face-Centred Cubic Arrangement. Observe in Fig 122a how 6 spheres surround each sphere. If you then placed horizontal layers, one above the other, so that upper spheres fit snugly into the lower dimpled areas, you would have a neat and particular orientation of the arrangement, very familiar to most people as the pyramid of oranges or oranges in closest packing. See Fig 122a.

A knowledge of Pyramidal Numbers of distinct height allowed armies to easily and infallibly count their stock of canon balls in predictable pyramidal heaps ready for the counting when the quarter master came to check the size of the stock.

Fig 123b on the following page, shows a mid-C19th ammunition store in Calcutta, India. Manuscripts shows that knowledge of such ammunition stacking began before 1600 AD.

Some mathematical professors state that knowledge of all these Figurate Numbers (the Squared, Triangular, Tetrahedral and other Polygonal Numbers) discussed in this book were known to the Babylonians more than 4,000 years ago!

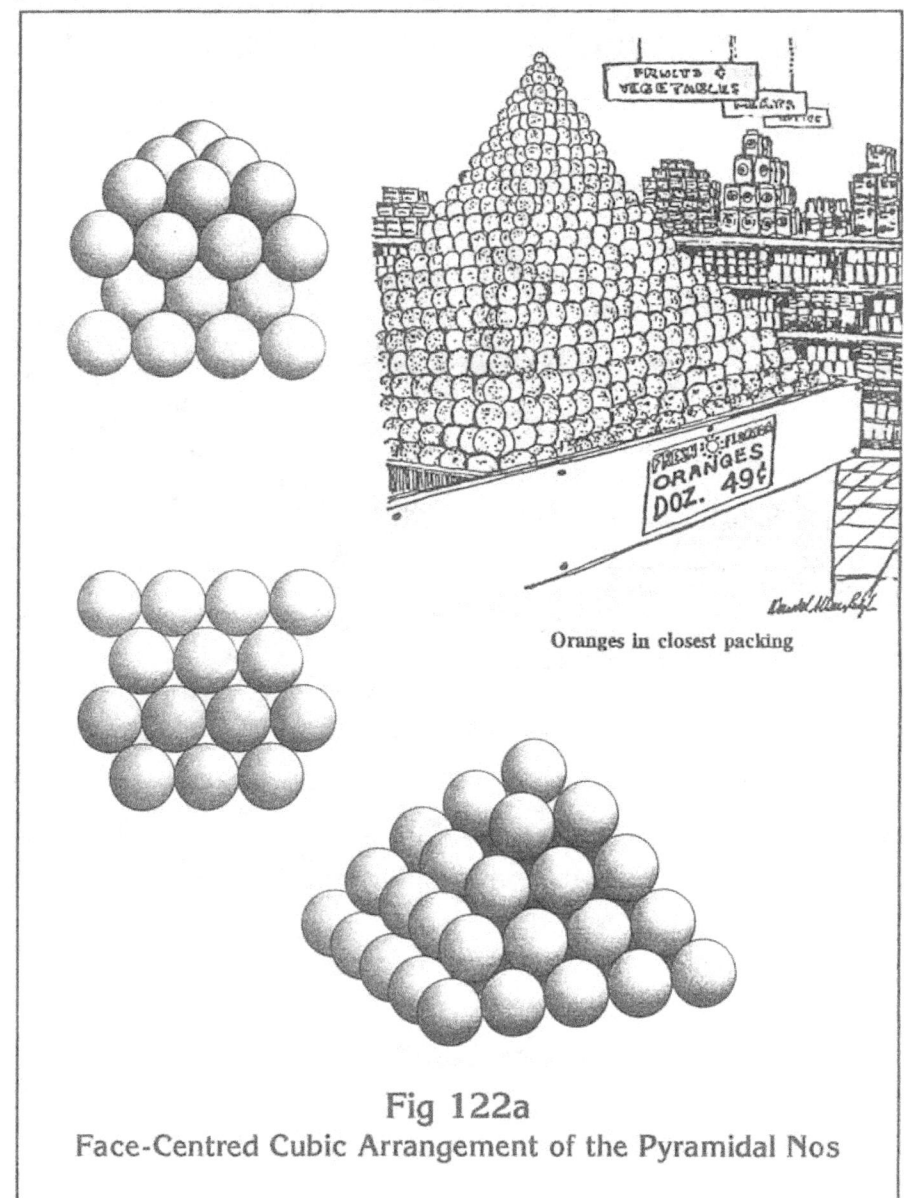

Oranges in closest packing

Fig 122a
Face-Centred Cubic Arrangement of the Pyramidal Nos

Fig 122b
The Stacking of Cannon Balls at an Ammunition Store in Calcutta, in the mid-C19th.

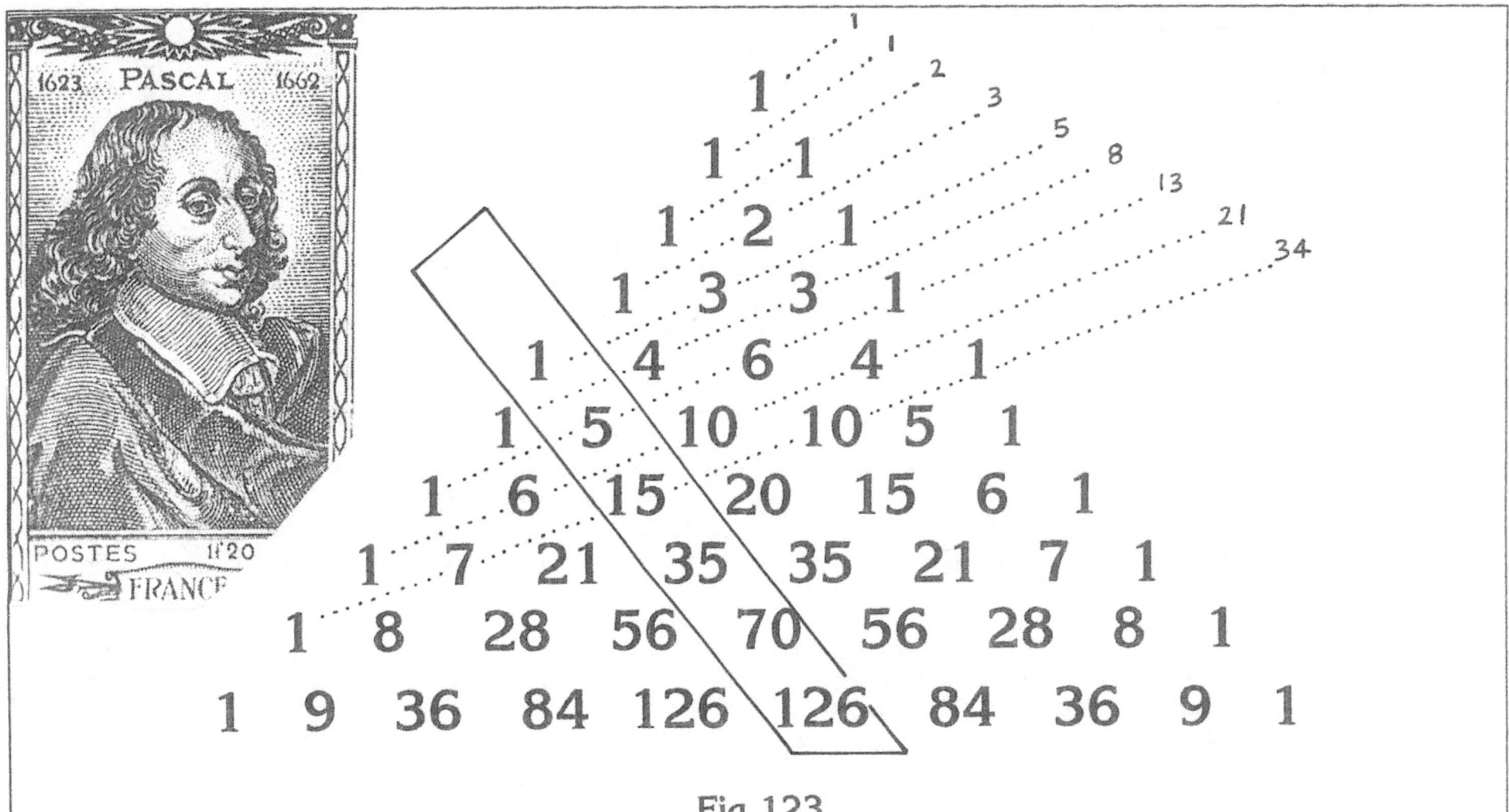

Fig 123

The Square Pyramid Numbers appear as the 5th Descending Diagonal Line of Pascal's/Halayudhas's Triangle. (Notice how the Fibonacci Numbers are formed by adding the digits along the dotted lines. See also Fig 115 for other Number Series appearing in this Triangle).

The mathematical formula to determine any Square Pyramidal Number is:

$$n \cdot (n + 1) \cdot (2n + 1) / 6$$

Thus the mental dialogue for the man in charge of counting the cannon-balls, wanting to know the final count for a pile 7 balls high, needed to know the 7th Pyramid Number utilising the memorised formula above:

"If I have a pile 7 high (n=7), I know I must duly multiply this number 7 by its successor (n+1) which is 8 and multiply this product, which is 7x8 or 56, by one more than the successor of its double (2n + 1) which is (2x7) + 1 which is 15, and then divide all of this by 6. This means (7x8x15) / 6 which calculates to 140 cannon-balls".

Which figurate group of Number Series do you expect to discover if you were to add the sum of 2 consecutive Tetrahedral Numbers?
(1, 4, 10, 20, 35, 56, etc).

eg: 4 + 10 = ...
    10 + 20 = ...
    20 + 35 = ...
    35 + 56 = ...

Mathematicians have known that:
The sum of 2 consecutive Tetrahedral Numbers is always a SQUARE PYRAMIDAL NUMBER.

What they don't know, and what has not yet appeared in print, to my knowledge and extensive research, are 2 unpublished (and 1 in print) simple observations about Pyramidal and Tetrahedral Numbers that generate the Magic Square Constants:

1.
THE SUM OF 3 CONSECUTIVE TETRAHEDRAL NUMBERS IS ALWAYS A MAGIC SQUARE CONSTANT.
(JAIN'S REVELATION)

This mathematical exposition was clearly shown and discussed in Fig 111.

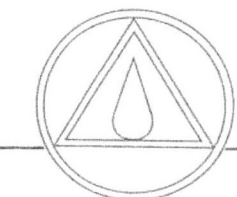

2. THE MAGIC SQUARE CONSTANTS ARE ALSO DERIVED BY PLACING THE 2 SERIES (THE TETRAHEDRAL AND THE SQUARE PYRAMIDAL NUMBERS) ONE ABOVE THE OTHER, BUT DISPLACED BY 1 UNIT, THEN ADD THE COLUMNS:

(JAIN'S REVELATION)

This is shown clearly below as Fig 124a:
(as this is one of two descriptions).

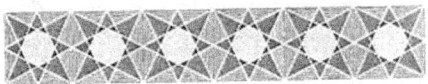

```
1   4   10   20   35   56            (Tetrahedral)
    +   +    +    +    +
    1   5   14   30   55   91        (Sq. Pyr.)
───────────────────────────────
1   5   15   34   65   111           M. Sq. Constants
```

Fig 124 a
Generation of the Magic Sq. Constants by Combination of the Tetra. and Sq. Pyramidal Nos.

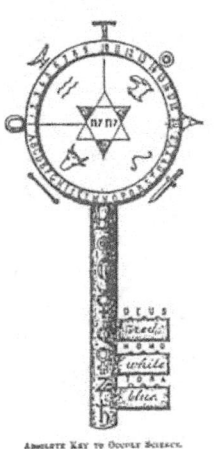

This simple and visual ability to perceive such unique formulae is one of the most ancient of keys

An alternate way of displaying this (Fig 124b) is to align the Tetrahedral and Square Pyramidal Numbers, directly one beneath the other, and add the diagonal relationships or sums to reveal the Magic Square Constants:

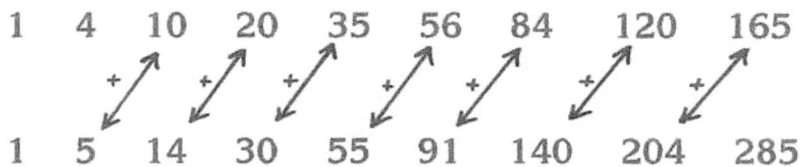

Fig 124 b
Magic Square Constants Revelation revealed as diagonal sums or pairs between the Tetrahedral and Square Pyramidal Numbers Series.

3.
THE MAGIC SQUARE CONSTANTS APPEAR AS A DISTINCT DESCENDING DIAGONAL OF THE CHARTED FIGURATE NUMBER SERIES.
(A.K.A GEOMETRIC AND POLYGONAL).

This has already been discussed in Fig 86. Although discovered or realised independently in 1984, this Magic Square Constant Revelation does appear in print, via the works of R. Brooks, 2001.

This section Part 2A now concludes and enters on a final and important revelation on "e".

The essence of this book is not that you need to know about Magic Square Constants, but more broadly than that, to be able to learn through visual discovery, patterns and relationships. Feminine Right Brain focus. I want this to be a book from which anyone who picks this up can merely look at the pictures and learn directly from the composition, order and arrangement of these **psycho-active**, sacred geometrical icons that speak their own universal language. What do you feel when you view the next 2 images?

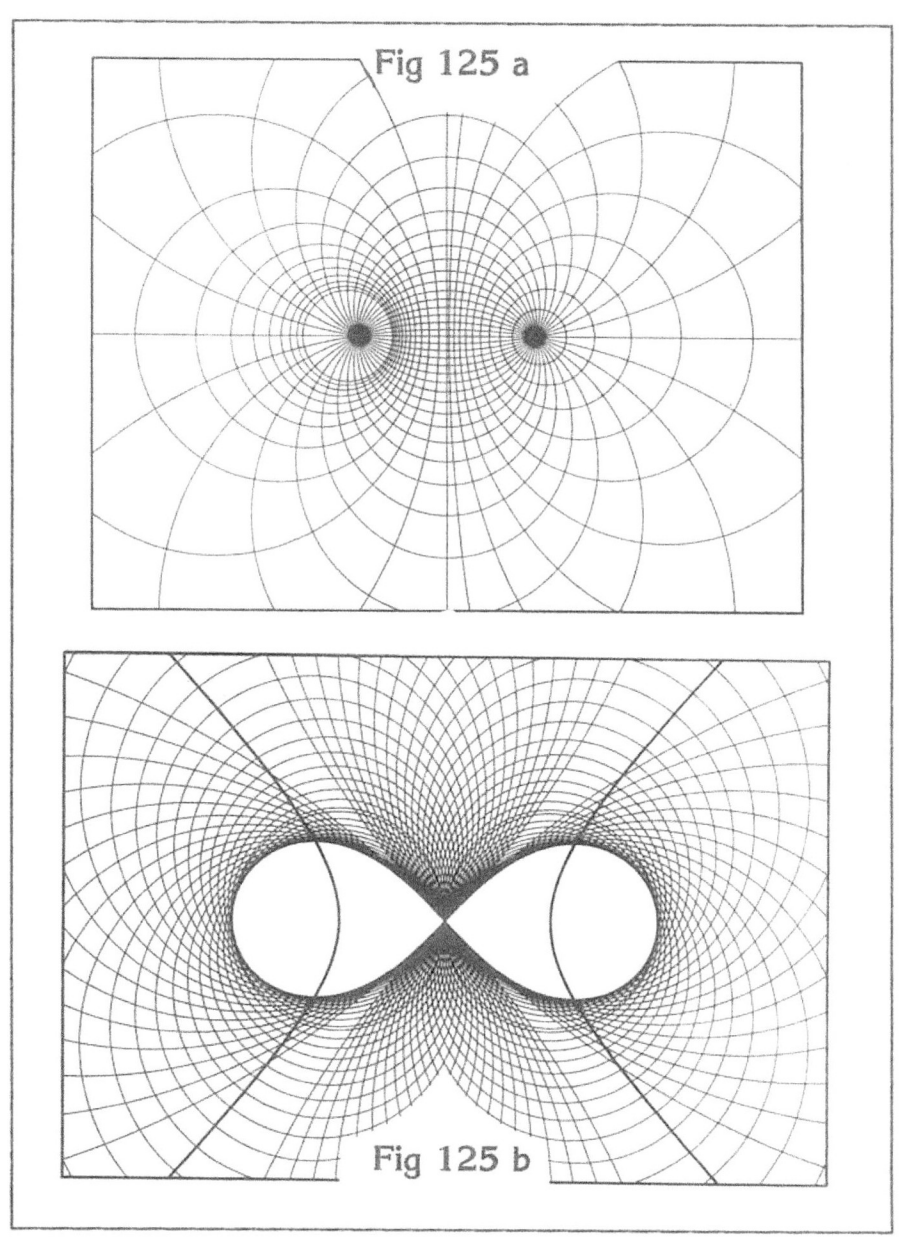

Fig 125 a

Fig 125 b

Is there a connection between the Square Pyramid Numbers and the following diagram: Fig 126 which is a square-based Pyramid in 3-Dimensions with sculptured segments to indicate a series of interpenetrating and successively smaller internal pyramids?

What is important here is a famous or well-known 2-D Indian diagram or yantra of **9 interlacing triangles** known as The Sri Chakra or **Sri Yantra** (shown below) that has been projected into 3-D (1996. Sculpture by Stan Tenen / Meru Foundation. www.meru.org/sriyantr.html).

Sacred Geometry is an ecstatic process and **transfigurational** when all squares are seen as cubes, circles as spheres, triangles as tetrahedrons, diamonds as octahedrons, pentagons as dodecahedrons and star-of-davids as star-tetrahedrons. That is, the ability to discern the 2-D shadows from their 3-D counterparts. The ability to be **In-The-Next-Die-Mansion** is the ability to be In-All-Dimensions Simultaneously.

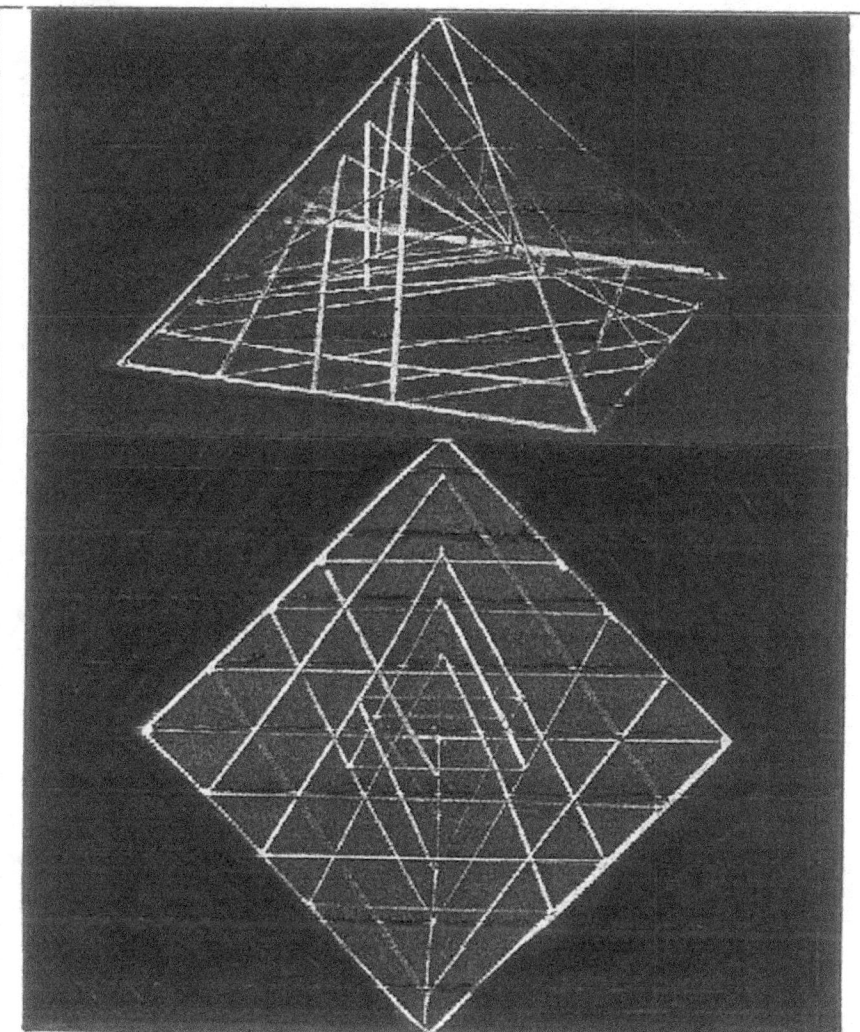

Fig 126
The Sri Yantra, the Mandala of Creation, Projected into 3-Dimensions. (Sculpture 1996. Stan Tenen)

CONCLUSION

Magic Squares are linked to the Tetrahedron God. The Magic Sums of the Magic Squares and Cubes constitute a Vibrational and Multi-Dimensional Jacob's Ladder. These are the New Frequencies for the New Earth embedded within our Bio-Crystal-Cells that carries immense trans-global information, knowledge and wisdom. The Tetrahedron's pure and stable Form and the Phi Spiral's ecosophical dance holds within their Geomorphic Memory Banks, the Galactic Mathematics of Magic Squareology and Cubeology that our children already KNOW, they do not need to learn by absorption and assimilation for they already are all unlimited Beings within their 12 helically stranded DNA. They already are the New Frequency, the Phi Codes for the Angelic repair of the damaged Earth Grid. Each day our many dimensional Bodies are being filled with a higher Light Quotient. You are advised not to study anymore, just Meditate on the Here and Now, every day, every morning to remember that your are just pure CONSCIOUSNESS.

The spin-rate of your chakras are the Magic Square Constants! DNA is not just a helix, the ends join to form a torus ring! The 3-D Hologram of the Pentacle is the Dodecahedron. This is mere and interesting intellectual knowledge. The Truth Is, without Love and Forgiveness of those who have hurt you, and No Blame, you will have learnt nothing.

Until you realise that no-one has done anything to you, that actually you have written the whole script, that you are the Creator of your Reality, then you have learnt nothing. Forget Sacred Geometry until you have done the emotional work. In fact, there is nothing to Forgive, for that implies that someone has done you wrong, but really there is no right or wrong. All things are allowed. There is no Judgement or Reaction. Just be a Witness. Learn to be Non-Reactive. Only then, when you have learnt to be Equanimous, the cells of your Being will teach you their matrix and their memory.

Visualise the desired outcomes you choose in your Life, now. When you meditate, don't ask, just visualise the desired outcome, eg if you have a child with arthritis, visualise that you are running on the beach with that child, in full health and fitness.

The problem with all the anti-war campaigns is that they feed the process of war. Everytime you say "ANTI", "NO", "DON'T" etc as in ANTI-WAR or "NO-WAR" the brain does not comprehend this negativity as our Spirit-Brain is Limitless Love, it only sees the words "WAR". When you tell a child to "DON'T DO THAT", they will do it. So drop all negative words and thoughts, and visualise your own God Self.

Part 3 B

An original and better way to express "e" in terms of 'Similar Powers', like

2^2 (2 squared or 2x2),

3^3 (3 cubed or 3x3x3),

4^4 (4 to the fourth power or 4x4x4x4)

5^5 (5 to the fifth power or 5x5x5x5x5)

exponential Number

=2.71828182845904523536028

$$e = 1 + \frac{1}{1!} + \frac{1}{2!} + \frac{1}{3!} + \frac{1}{4!} + \frac{1}{5!} + \frac{1}{6!} + \frac{1}{7!} + \frac{1}{8!}$$

Fig 127
The simplest known approximation for "e"
by the Swiss Mathematician, Euler (1707 – 1783)

(nb: the above symbol "!" is pronounced "factorial" and means multiplication of numbers in consecutive order. Thus
5! = 1 x 2 x 3 x 4 x 5)

Euler (Switzerland B267)

WHAT IS "e" ?

• The Exponential Number "e" is more closely related to Human Affairs and Growth than Pi.

• "e" is known as a constant, like pi (Π) and phi (Φ) and "i" (the square root of negative one), in that they do not distinctly solve algebraic equations, are non-repeating, non-terminating, and generated by an infinite series, like the sum of the fractional factorials above in Fig 127. Mathematicians are surprised by this entity called "e", which is most famously used for compound interest in the well known formula

$(1 + 1/n)^n$ where one would expect an infinite return, but what happens is that it actually reaches a limit called the exponential function of 2.71818 etc.

Therefore it is defined as the limit of an infinite series.

- Other areas where e is used is in Navigational Tables, Radioactive Decay, Maths and Applied Science, Economics, Theory of Probability.

- The Exponential Function $y = e^x$ relates to the behaviour of <u>growing things</u>. It is unique in that it is the only function of "x" with a rate of change, with respect to "x", that is equal to the function itself. As a graph it is expressed as Fig 128 below:

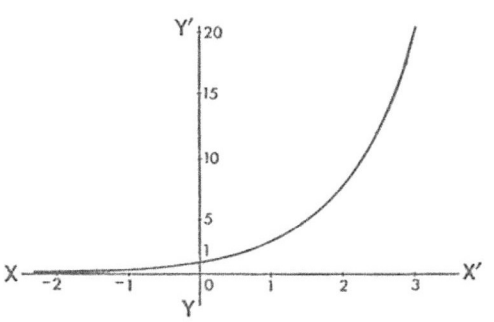

Fig 128

Graph of $y = e^x$

- A "Function" has variable quantities where a change in one implies some change in the other. eg: the cost of quantity of tofu or sushi is a function of its weight. The speed of a water-powered car is also a function of the amount of water it consumes as non-polluting fuel. The amount of perspiration that you give off is a direct function of the temperature. Can you can think of any other examples, like a projectile in full flight, a current flowing through its wire, the growth of a human or bacterial population, of a tree, of an omeba. This is therefore the peculiar meaning of all organic processes as symbolised by the graph of Fig 128, that <u>the rate of growth is proportional to its state of growth</u>, which simply means the bigger some entity is, the faster its colony will grow as in the population of China, the bigger it is the faster it grows.

- "e" is unique and valuable in many branches of maths especially in <u>Calculus</u> which solves problems of moving objects by measuring the area under its plotted graph. Mathematicians around Euler's time quickly realised that <u>Logarithms</u> which are based on the powers of 10 (eg: 1000 = 10 to the power of 3, and therefore its logarithm is 3) were inefficient. They were good in that large multiplications and divisions were solved by mere addition and subtraction, but they learnt that e = 2.718 was a preferred base for the expression of powers, better than 10 and related to natural occurences, that "e" became known as the <u>Natural Base of Logarithms (or Log 'e')</u>. and is used today by all financiers and scientists in solving problems involving rates of change.

- There are many mathematicians, like myself, who are enamoured by Prime Numbers. One of Gauss's great insights was to discover that the distribution of prime numbers related to 'e'. As the Prime Number Series grows and gets bigger and thinned out, their density is inversely proportional to 'e', the natural logarithm. What this means is that if a mathematician wanted to explore the density of prime numbers around the number 100, he would inspect that there are 9 prime numbers between 75 and 125 which is 1 prime in every 5.5 numbers and he already knows that the nat. log of 100 is 4.6 which is very close to the 5.5 number. This means he now has a reliable tool to predict how many primes to expect in the range of a thousand or a million. Such formulae are indeed portals In The Next Dimension. It is my intention to offer another original Formula equally as beautiful and simple as that revealed by Euler in Fig 127.

If 'e' appears mysteriously in the Prime Number Sequence, does it therefore appear everywhere. Is it omnipresent?

Here now is the new 'e' formula, again to my knowledge, hitherto unpublished. It is very beautiful as it incorportates a passing through all the dimensions which will be expressed as the rising Powers of $n, n^2, n^3, n^4, n^5, n^6, n^7, n^8, n^9$ etc

The following is a veritable expression of the important Vedic Maths Sutra: "BY ONE MORE THAN THE PENULTIMATE DIGIT" and therefore the exemplification of IN THE NEXT DIMENSION, the 17th, and Lost Sutra.

I will give an example now, then the algebraic formula, then explain how simple it really is:
(the symbol here '≈' means 'approximately equal').

$$e \approx \frac{5 \times 5 \times 5 \times 5 \times 5}{4 \times 4 \times 4 \times 4} - \frac{4 \times 4 \times 4 \times 4}{3 \times 3 \times 3}$$

which is read as "5 to the 5th Power (5^5 or 5^5) divided by 4 to the 4th Power (4^4 or 4^4) minus 4 to the 4th Power divided by 3 to the 3rd Power (3^3 or 3^3) equals approximately to 2.725549769.

If this formula holds true we would expect that as we increased the numbers into the hundreds and thousands we would expect to arrive at exactly 2.71828 etc, as is the case and will now be shown.

What is important here is that this formula to

be given was not worked out mathematically, it was realised to me via pure INTUITION. I merely had to check the numbers to confirm that my INNER KNOWING was correct. This is therefore another meaning of the Sutra: IN THE NEXT DIMENSION.

As you can see I am deliberately hesitating to give you the algebraic formula as I know that many wounded students are baulked and frightened when they see such language and instantly shut down. When you study the Eastern Vedic Mathematics you will clearly realise why Western Mathematics is clumsy and inappropriate and needs to be updated. Here's the algebra, just embrace it and read on:

$$\left[\frac{(n+2)^{(n+2)}}{(n+1)^{(n+1)}}\right] - \left[\frac{(n+1)^{(n+1)}}{(n)^{n}}\right]$$

Fig 129
JAIN'S ORIGINAL 'e' FORMULA,
THE NUMBER OF GROWTH AND DECAY,
BASED ON VEDIC MATHEMATICAL SUTRAS
AND THE RISING 'SIMILAR POWERS'.

HOW TO UNDERSTAND THE FORMULA:

1) – Choose any 3 consecutive numbers, like 142, 143, 144
and call them the 1st, 2nd and the 3rd respectively.
2) – Raise each number to its own power.
(This is what I mean by "similar powers" a term that I have invented).
ie: 141^{141}, 142^{142}, 143^{143}
3) – Divide the 3rd by the 2nd ...... call this [A].
   Divide the 2nd by the 1st ...... call this [B].
4) – Subtract [B] from [A].
5) – You now have 'e'.

6) – In the Formula of Fig 129, these 3 consecutive numbers are expressed as:
$(n)^{n}$ for 142^{142}
$(n+1)^{(n+1)}$ for 143^{143}
$(n+2)^{(n+2)}$ for 144^{144}

7) – The answer for this particular example using the above 3 numbers: 141, 142, 143 comes down to
= 387.354365734 minus 384.636078288
= **2.71828**74456 correct to 10 DP (Decimal Places) but the official 'e' is **2.71828**1828459045235360 etc, so what we have here is a correct answer to 5 DP as shown in bold and underlined: 2.71828.

This was first hand-written and therefore © on 27/2/2003, Full-Moon in Virgo, Mullumbimby Creek.

I now invite any readers to check this with their advanced computers with larger and larger 3 consecutive numbers like: 998, 999 and 1,000 and hopefully improve on extending the correct decimal places for 'e' and thus proving that this paradoxical **"Infinite or Unlimited Series With a Limit"** is a highly intelligent, sophisticated and economical formula that Nature's Realm of Growth and Decay has chosen as its Mathematical Ambassador.

The 5 most important symbols in mathematics are 1, 0, *i*, Π and e (where "i" = the square root of −1, another subject as deep and interesting as 'e'), can all be combined into one single formula that has fascinated mathematicians, scientists and mystics for 200 years since Euler popularised it! We still use the same college-level notation that Euler developed, in fact he is regarded as the most successful notation builder of all times, even popularising the symbols for "e", "*i*" and "Π". His now famous formula was:

$$e^{\Pi.i} + 1 = 0$$

which he meditated upon in his last 17 years of life that he spent in total darkness being blind. He published over 500 books and papers in his life time.

© 1964 United Feature Syndicate, Inc.

APPENDIX

This will be in 3 Sections:

1: Why the Number 9 is Uniquely Inter-Dimensional?

2: An Advanced EARTH-HEART Meditation.

3: Platonic Solid and Phi Spiral Diagrams for general FotoCopy use, as a Teacher's Resource Material.

APPENDIX

1.
Why the Number 9 is uniquely Inter-Dimensional?

The Number 9 is the **only** Square Number that is the sum of 2 consecutive Cubic Numbers:

$$3^2 = 1^3 + 2^3$$

9 has the unique ability to move between the Dimensions, to magically dematerialise from the 3-Dimensional Cubic Realm ($1^3 + 2^3$) and re-appear as a 2-Dimensional plane figure (3^2) and vice-versa.

9 (as 3^2) is therefore able to communicate effectively with its next Higher World or Cubic Reality. That is why 9 is the Key to Vedic Mathematics as it knows how to do its magic
IN THE NEXT DIMENSION.
(See my 2 books and Video:
1. The MAGIC of 9 IN VEDIC MATHEMATICS.
2. The VEDIC MATHEMATICS CURRICULUM for the GLOBAL SCHOOL. Part 1 : Digital Sums.
3. VIDEO: VEDIC MATHEMATICS FOR THE NEW MILLENNIUM. Part 1: The Magic Of Nine.

2.
AN ADVANCED EARTH-HEART MEDITATION
(Guided and Illustrated by Jain)
(Standing Barefoot and/or Naked on a Rock)

- Begin by moving energy through your Pranic Tube, as illustrated in Fig 130a. It runs from the top of your crown (fontanelle, the soft spot on a baby's crown) and thru your perineum (the contractible love muscle located between your penis/vagina and anus).
- Hold your thumb and index finger joined, like in the classical Indian hand mudra, or just face your open palms skyward.
- Breathe in via your left foot, being conscious or aware that the Left side of the body "RECEIVES" energy, in contrast to the Right side of the body that "GIVES OUT" energy. Direct the rising Earth energy to the spot 2 finger breadths below your navel, that which the Chinese call Dan Tien, and is also picturised as a 'Cauldron'. Allow this to pass thru the Navel chakra (literally, "wheel" in Sanskrit) and down your right leg and out of the center of your right foot and back to Source which is the Earth's Crystalline Core. Feel that there is a continuous loop between

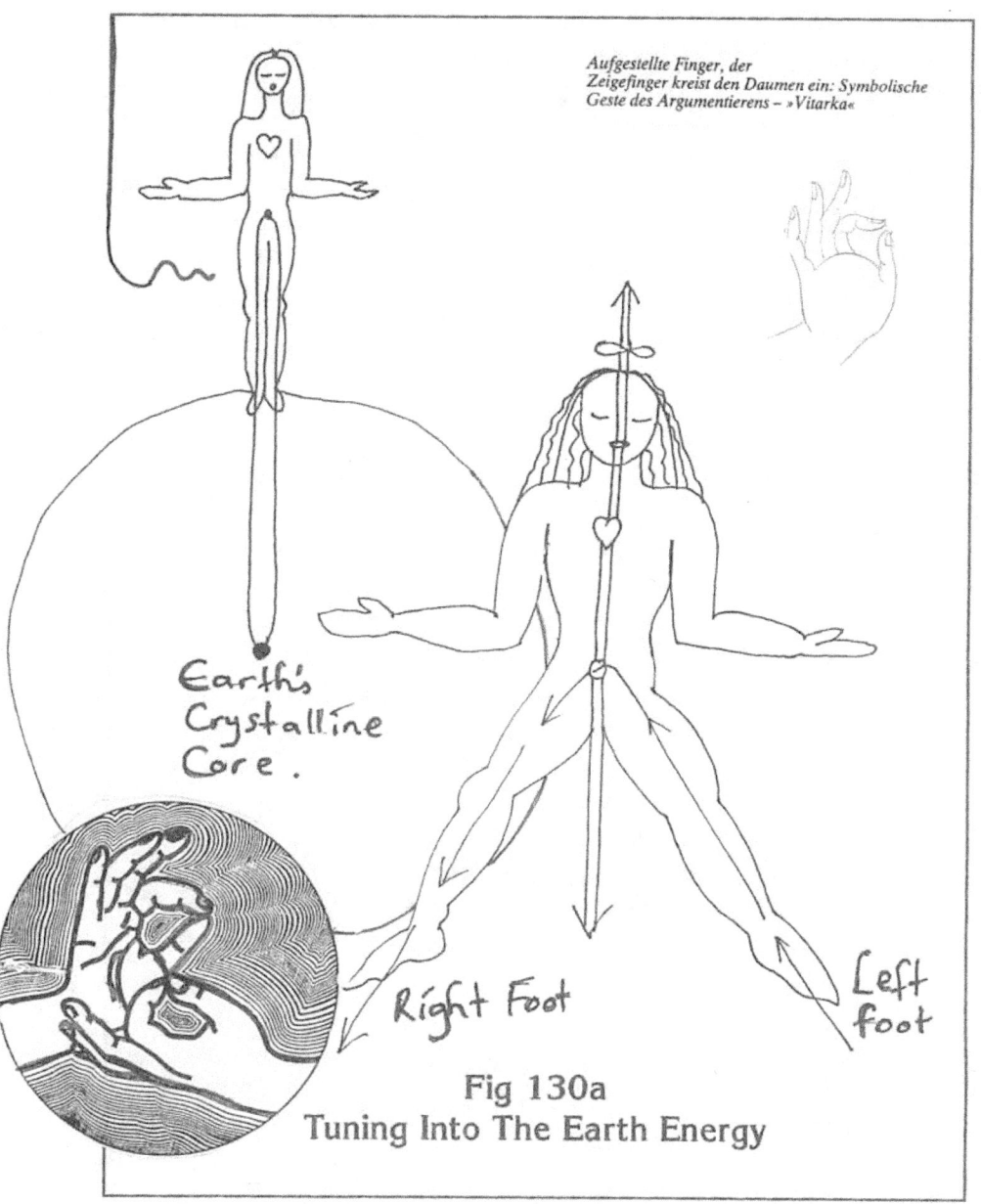

Fig 130a
Tuning Into The Earth Energy

you and Earth, giving Gratitude that you are Remembering the Truth of Earth's History and Memory. In effect, you are <u>Tuning In From Your Centre To Its Centre</u>.

- Breathe in via your receptive Left Hand, both arms upraised like a funnel, to receive the Sky or Cosmic Energy, as illustrated in Fig 130b. This Golden Light runs thru your Heart Chakra (Anahata) and out via your radiating Right Hand, back to celestial Source. This declares you as a Divine Conduit for The All-That-Is. It gives you a feeling of certainty, in this Now-ness of who you really are, where you are going and what your Divine Plan/Plane/Planet on Earth is. It will download the Truth from amidst all the confusing reports of the so called galactic creation myths/fairytales whether you are a mere DNA alien experiment that went wrong a million years ago, or whether planet Earth is really a harvested DNA Library/Gene-Splicing-Bank under the control of Master Geneticists who shape-shifted here, or whether you are truely **Goddessence** in the Making etc. You may visualize the Path of Energy coming in and out as a gigantic Love Heart. Just remember that there is no ambition, no stress, <u>nowhere</u> to go, that in the very word: NOWHERE is really **<u>NOW HERE</u>**.

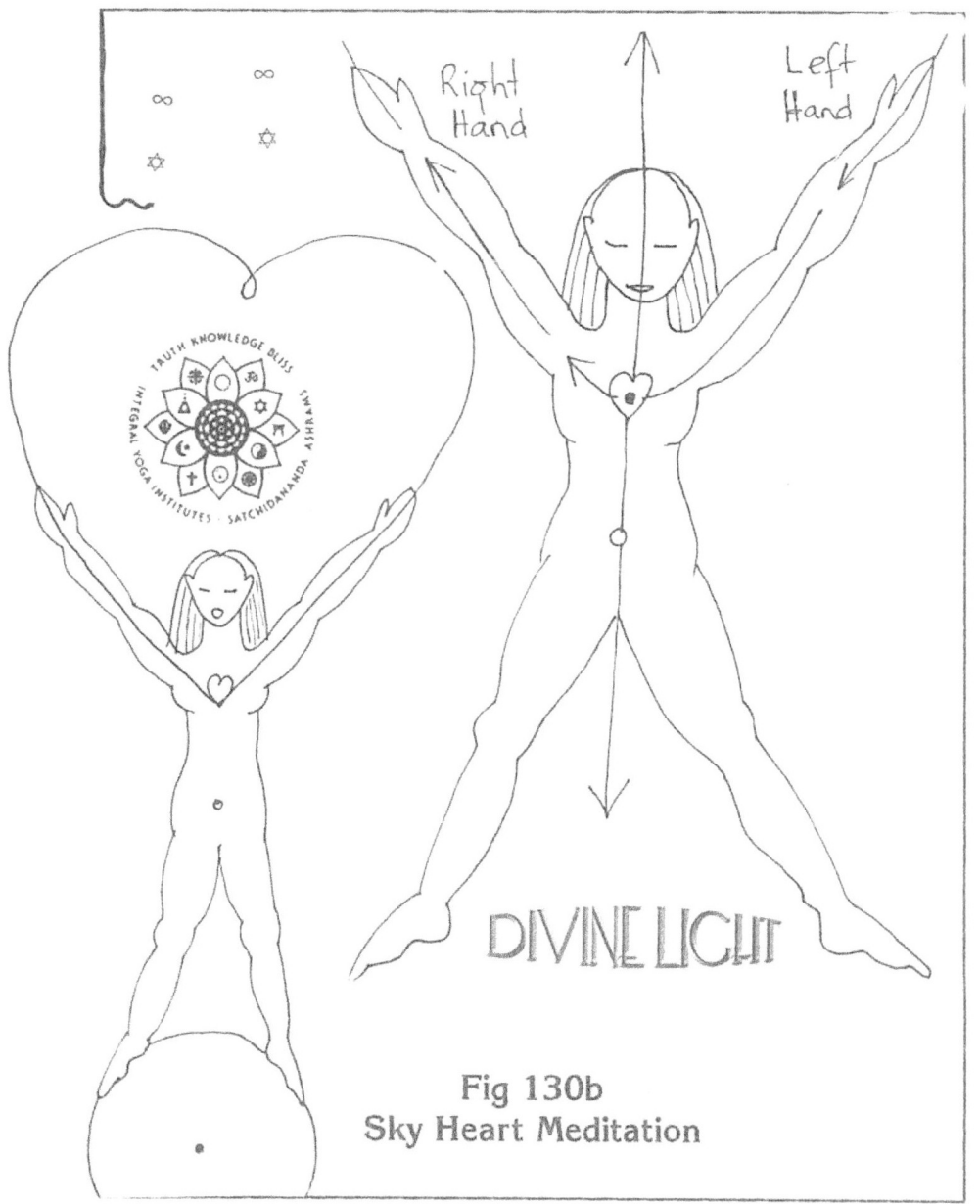

Fig 130b
Sky Heart Meditation

We now proceed to combine the two separate meditations and make them as One Advanced Meditation. To bring Heaven on Earth. Remember this acronym:

HOME = Heaven On Mother Earth.

Another clever way of expressing this is shown on the front cover logo of most of my books (which is being used by people all over the world now): If you necklace the word EARTH many times:

EARTH EARTH EARTH EARTH EARTH

and it now reads HEART HEART HEART many times:

EART**HEART**HEART**HEART**HEARTH

• The advanced EarthHeart Meditation is a combination of the two previous parts which persuades the accumulated energy to link in the form of the Eternity Symbol whose centre is the Heart Chakra. (Remember that the Eternity Symbol, like the double Sine Wave, viewed as the Reclined Figure/Number "8" is really the cross-section of the Tube Torus Doughnut, best-shown in **Fig 57**, and also in Figs 14a, 35, 59, 70 and 71).

This creates Heaven On Earth. A Oneness with All Creation. A Unity With All Cultures. All false concepts of Duality and Separation from God/dess now

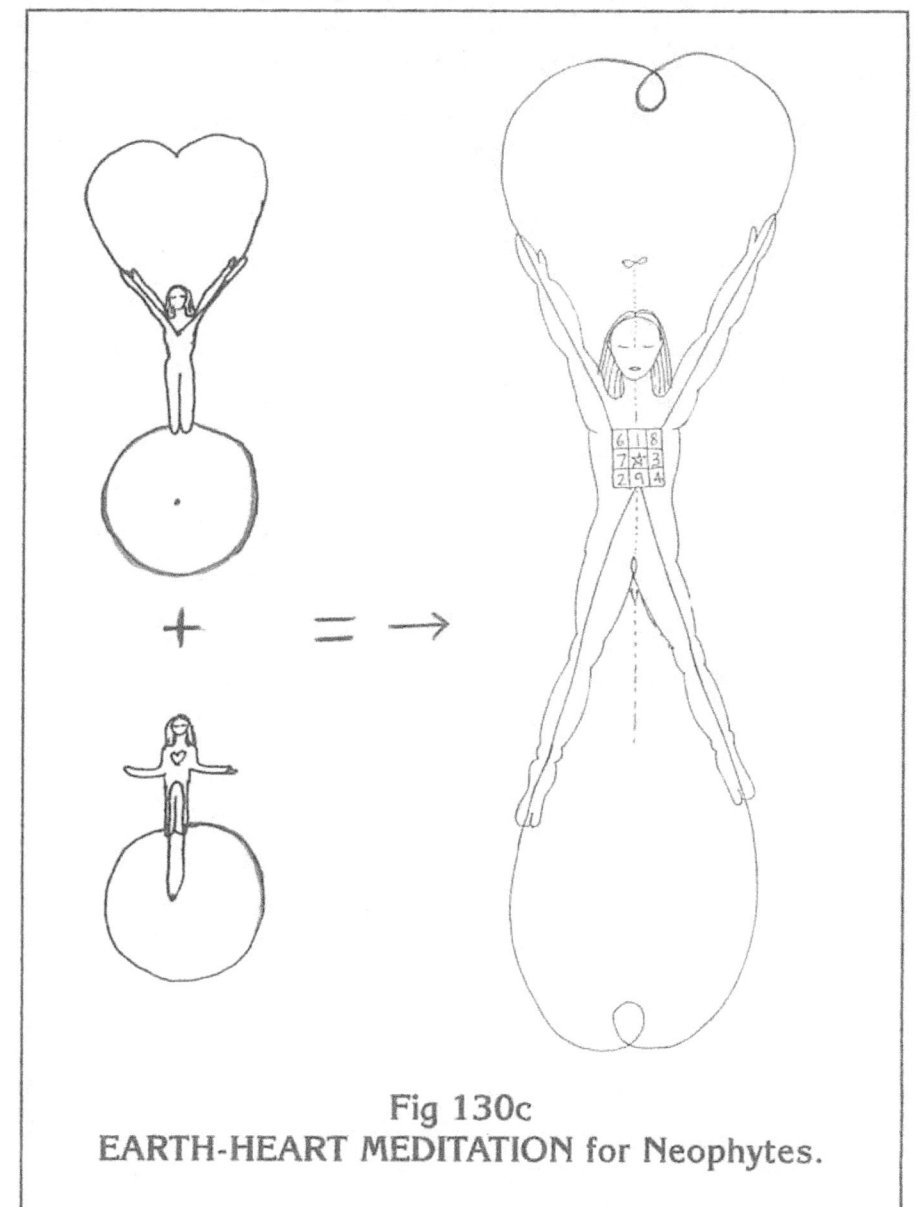

Fig 130c
EARTH-HEART MEDITATION for Neophytes.

d i s s o l v e .

The Advanced Student or Neophyte can forego the previous two steps and begin with this combined Earth-Heart synthesis. The preferred choice of symbolism here is to visualise the The Magic Square of 3 x 3 or Lo-Shu of 9 Numbers or Cells reMinding you of your OmniPresence or existence simultaneously in the 9 Dimensions with 3 Physical Bodies (physical, emotional and mental). (This is the reason why the page: "Why the Number 9 is Uniquely Inter-Dimensional?" was listed at the beginning of this Appendix).

As shown in Figs 48 to 52, the Magic Square of 3 x 3 actually creates the 3-Dimensional Form of the Star of David known as the Star TetraHedron or Stella Octangula which is your veritable MerKaBah or Spiritual Vehicle of Light that allows you to transcend both Space and Time and BE wherever the Power of your Thoughts and Love are willed.

At the end of your meditation, hold your hands as depicted in Fig 130d, which honours the union of your celestial and physical body TEMPLES.
In Love and Service,
Jain. 19/9/1996, Royal National Park, Sydney, during a 2 week juice (fast)/slow.

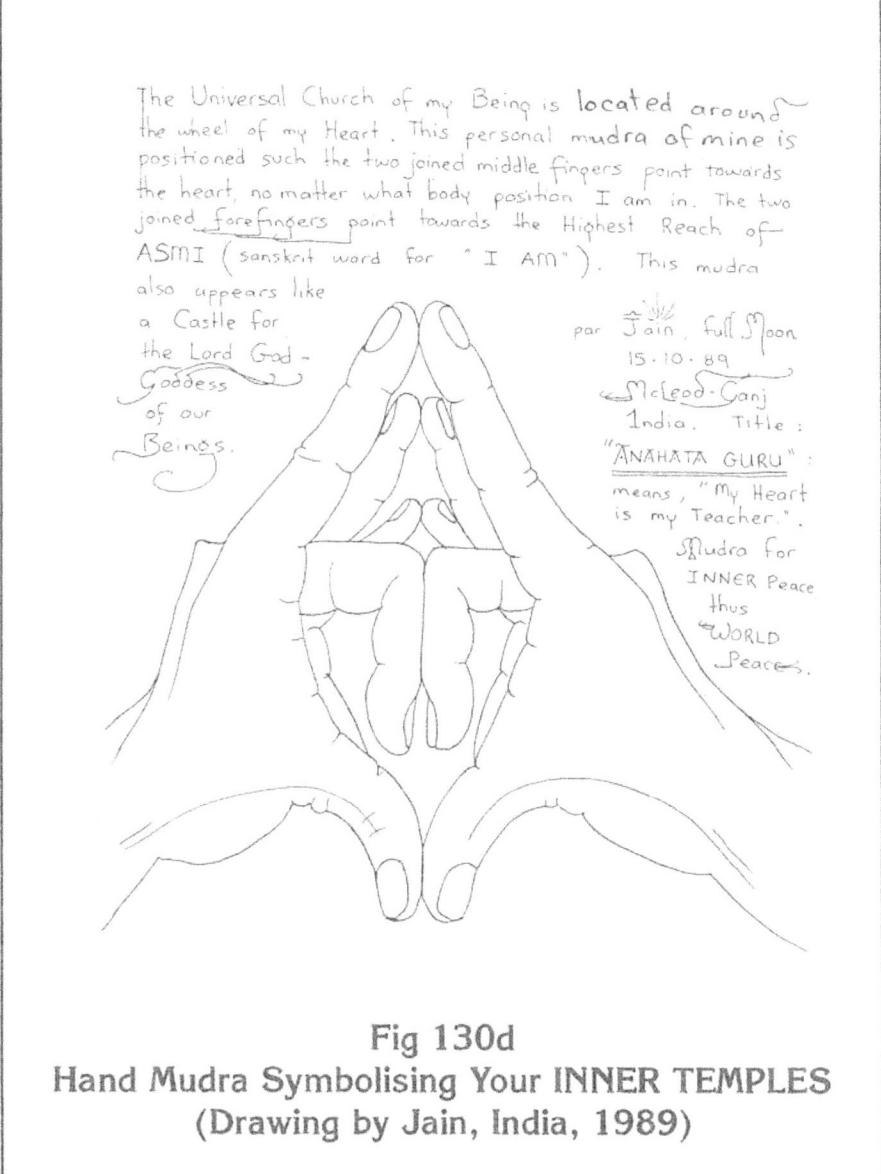

Fig 130d
Hand Mudra Symbolising Your INNER TEMPLES
(Drawing by Jain, India, 1989)

SUMMARY OF THE 14 STEPS OF THE EARTH-HEART MEDITATION.

The Earth-Heart Meditation is a One Breath Meditation designed as a 14 step process for a full one breath. It consists of 2 x 7 "S"-shaped pranic movements shown as Figs 130e and 130f.

To begin, it helps to bend your knees a little and tilt your coccyx (tailbone) forward, so as to straighten the spine.

Follow the first 7 steps of fig 130e which is really an "S"-shape or half an "8"-shape. All on the In-Breath. The Centre Point (4) is your choice of the Heart or Solar Plexus Chakras.

Receive Sky Energy (1) via your left hand (2) and down (3) to centre (4) crossing over and down the right thigh (5), and out via the right foot (6) into the Crystalline Core of the Earth (7).

Fig 130f shows the continuation of this: Receive the collected Earth Energy via the left foot (9) up the right thigh (10) to centre (11) crossing over to the right arm (12) and out via the right hand (13) and radiating out this Love to all Beings of the many Galaxies (14).

Symbolically, the 14 steps activate the vibration of the Number or Spirit of 14 which occupies the central cell of the holographic Magic Cube of 3x3x3.

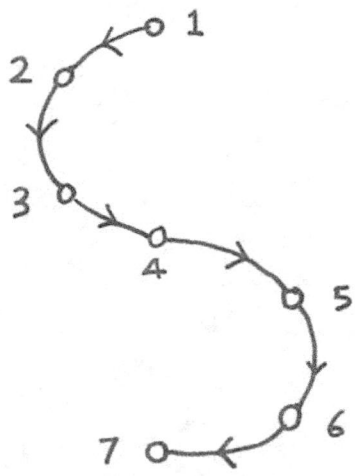

Fig 130e (Steps 1 to 7 of In-Breath)

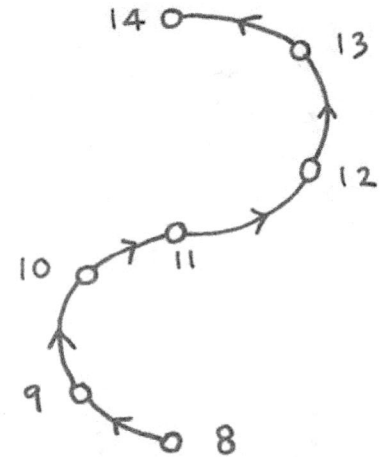

Fig 130f (Steps 8 to 14 of Out-Breath)

EARTH-HEART

The Earth-Heart Meditation is basically the natural flowing figure of 8 movements of Breath.(The figure of 8, on its side, like the infinity symbol is the sine-wave cross-section of the Torus Doughnut shape).

At the end of your meditation you may want to simplify the 14 steps mentioned in Figs 130e and f, relax from the conscious tracing of the "8"-shape and simplify the whole meditation to a single line of the Pranic Tube (Fig 130g).

- Visualise yourself standing in a gi gantic toroidal field, doughnuts nesting within bigger doughnuts, like you are the sun centre of a whole solar system.
- On the In-Breath, draw in Prana Life Force thru your crown and perineum simultaneously, and allow these energies to meet at the Heart.
- On the Out-Breath, radiate all this Love to all Humanity.

Very simple and potent. Repeat this as many times as needed.

- Only do these meditations when you are enjoying them, not out of habit or ritual. Sometimes, when the opportunity is available, stand naked on a rock under the direct sun or sunlight filtered thru the trees.

Be cautious of complex sacred geometrical techniques that utilise the Platonic and Archimedean Solids. If it is not simple, avoid it.

Avoid chanting any words of power or chants or mantras that you do not understand.

- Each Day, give **GRATITUDE** for all the **ABUNDANCE** that you already have.

Fig 130g
Simplified Earth-Heart Pranic Tube Meditation

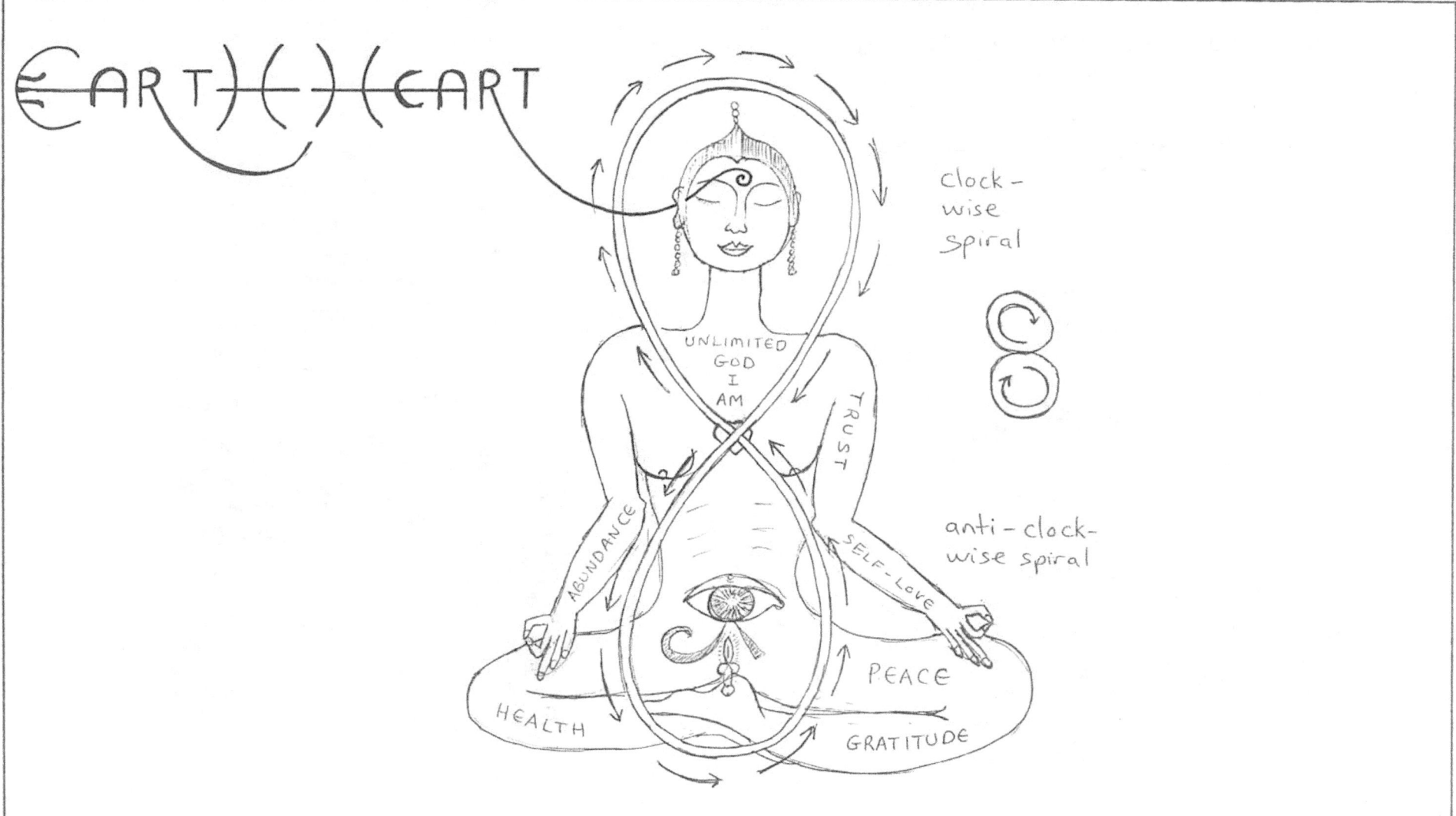

Fig 130h EARTH–HEART MEDITATION IN SEATED POSTURE. by Jain, 2003
UNIFICATION OF COUNTER-ROTATING SPIRALS MEETING IN HEART AND MOVING THRU BREATH

3.
Platonic Solid and Phi Spiral Diagrams for general FotoCopy use, as a Teacher's Resource Material.

Fig 131a
Annie Besant, Micro-Psi Artist who collaborated with Leadbeater

Fig 131b
Charles Leadbeater, author of "OCCULT CHEMISTRY" who accurately drew 'Micro-Psi Images" of all the Elements + Isotopes of the Periodic Table 100 years ago! associated with the Theosophical Societies in London and India

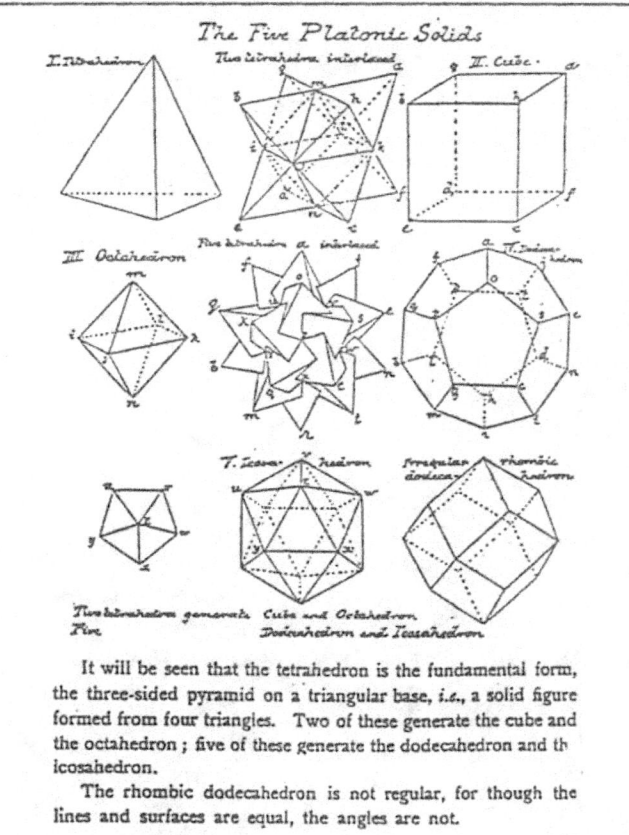

Fig 132a

Reproduction of page 15 from the book "Occult Chemistry" pub. 1909. It clearly declares that the Platonic Solids are the building blocks of chemical elements and some are generated from interpenetrating tetrahedra in special ways. This is exactly what physicists have found to be the case with the shells of nucleons in magic nuclei. (from Chris Illert's book)

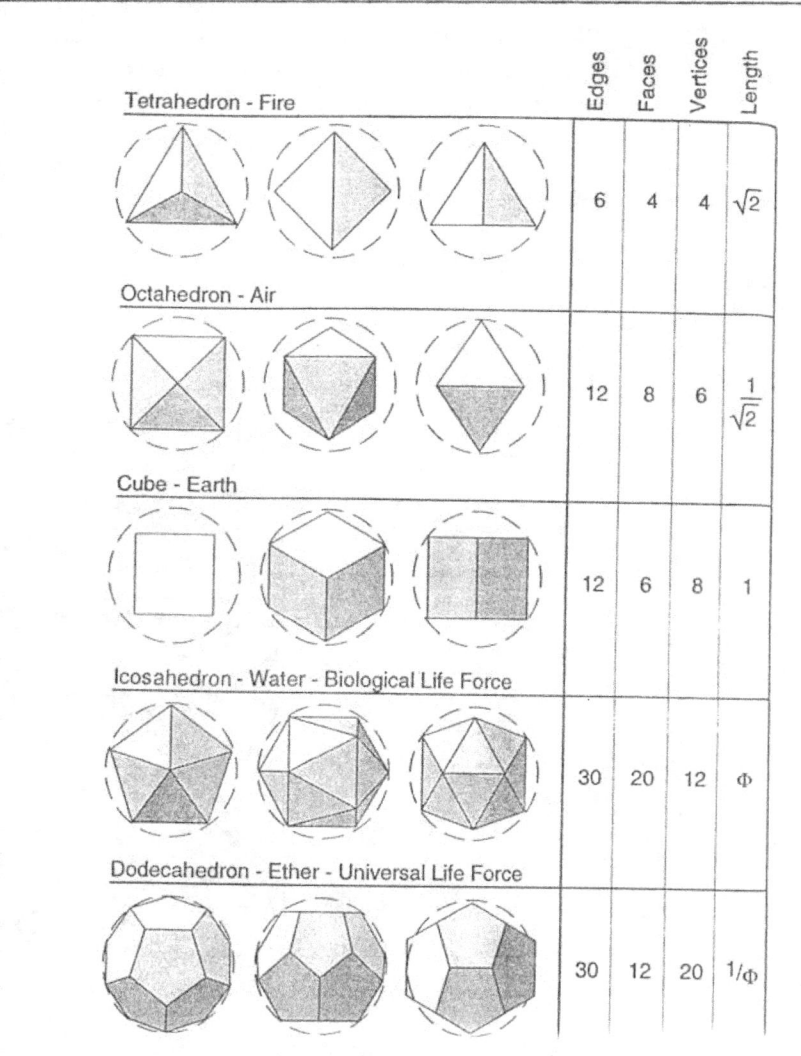

	Edges	Faces	Vertices	Length
Tetrahedron - Fire	6	4	4	$\sqrt{2}$
Octahedron - Air	12	8	6	$\frac{1}{\sqrt{2}}$
Cube - Earth	12	6	8	1
Icosahedron - Water - Biological Life Force	30	20	12	Φ
Dodecahedron - Ether - Universal Life Force	30	12	20	$1/\Phi$

Fig 133b
The 5 Platonic Solids: equal angles, lengths, faces

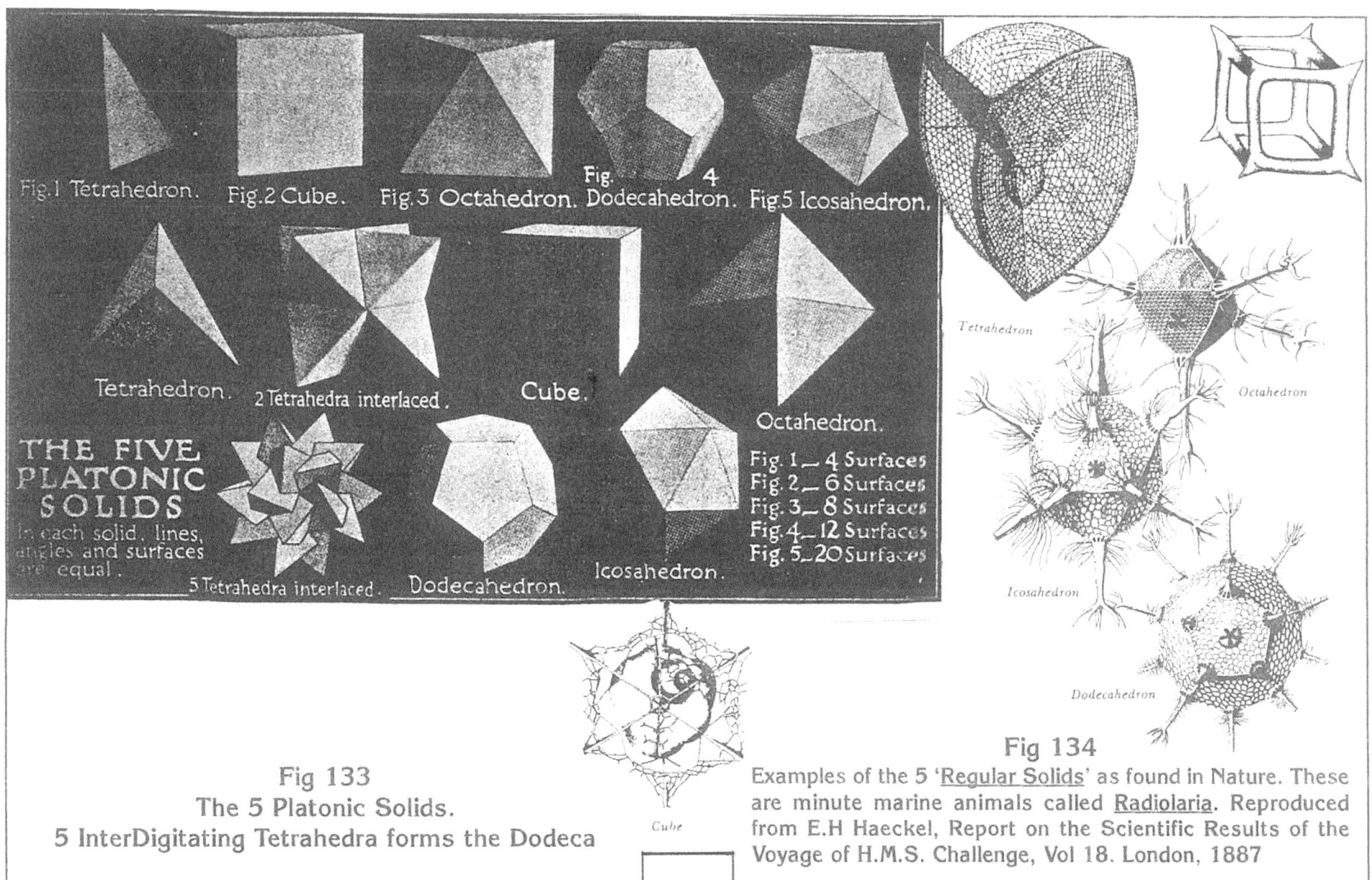

Fig 133
The 5 Platonic Solids.
5 InterDigitating Tetrahedra forms the Dodeca

Fig 134
Examples of the 5 'Regular Solids' as found in Nature. These are minute marine animals called Radiolaria. Reproduced from E.H Haeckel, Report on the Scientific Results of the Voyage of H.M.S. Challenge, Vol 18. London, 1887

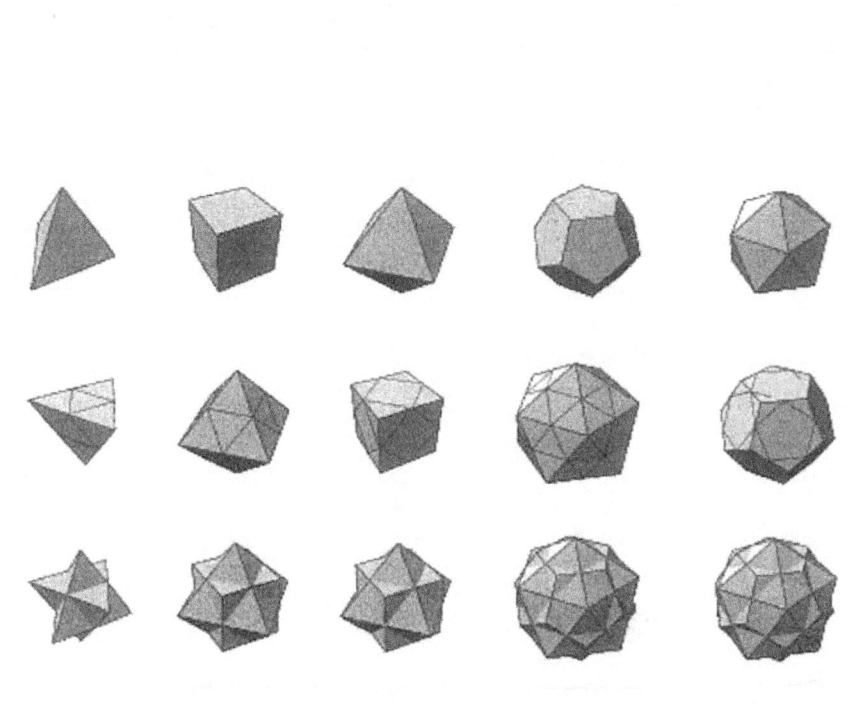

Fig 135

For every polyhedron there is another internal and complementary shape called its <u>DUAL</u>, achieved by plotting the centre of the external polygonal faces and joining them to form new vertices. This duality principle is also called Reciprocation. (Interesting shapes occur when you plot the midpoint of the edges and join those).

The top row above is the Platonic Solid. The middle row is its corresponding Dual, and the bottom row is combining the two top rows to form attractive shapes called the Polyhedron Compounds.

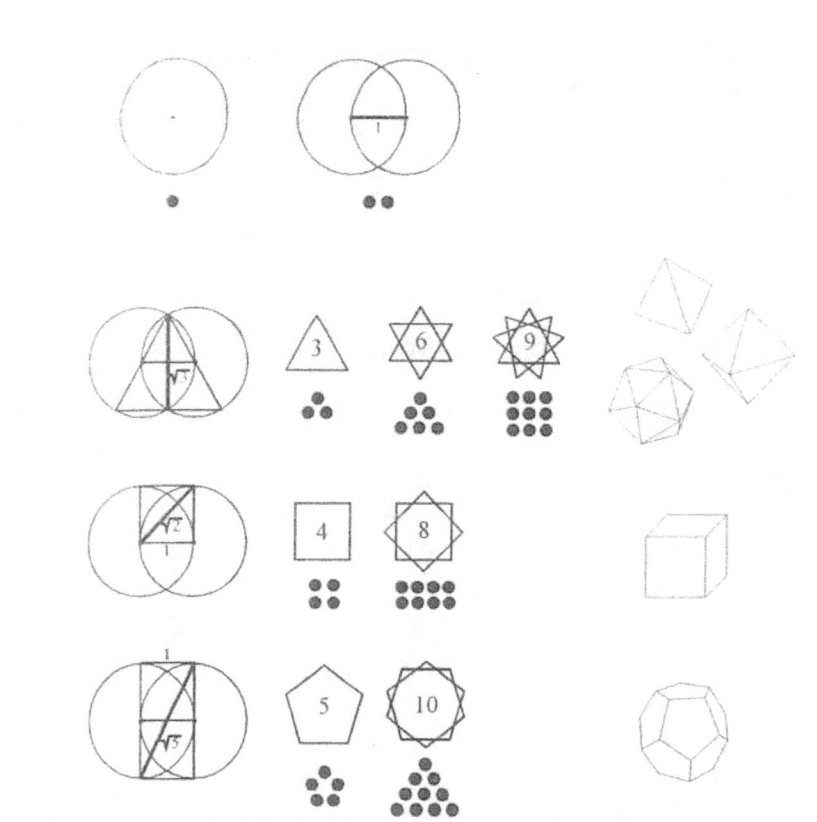

Fig 136

The intersection of 2 circles going through each other's centres, known as the <u>VESICA PISCIS</u>, contains all the vital proportions in Sacred Geometry that generates the polygons expressed in Nature (except for the septagon) and the volumes of the Platonic Solids

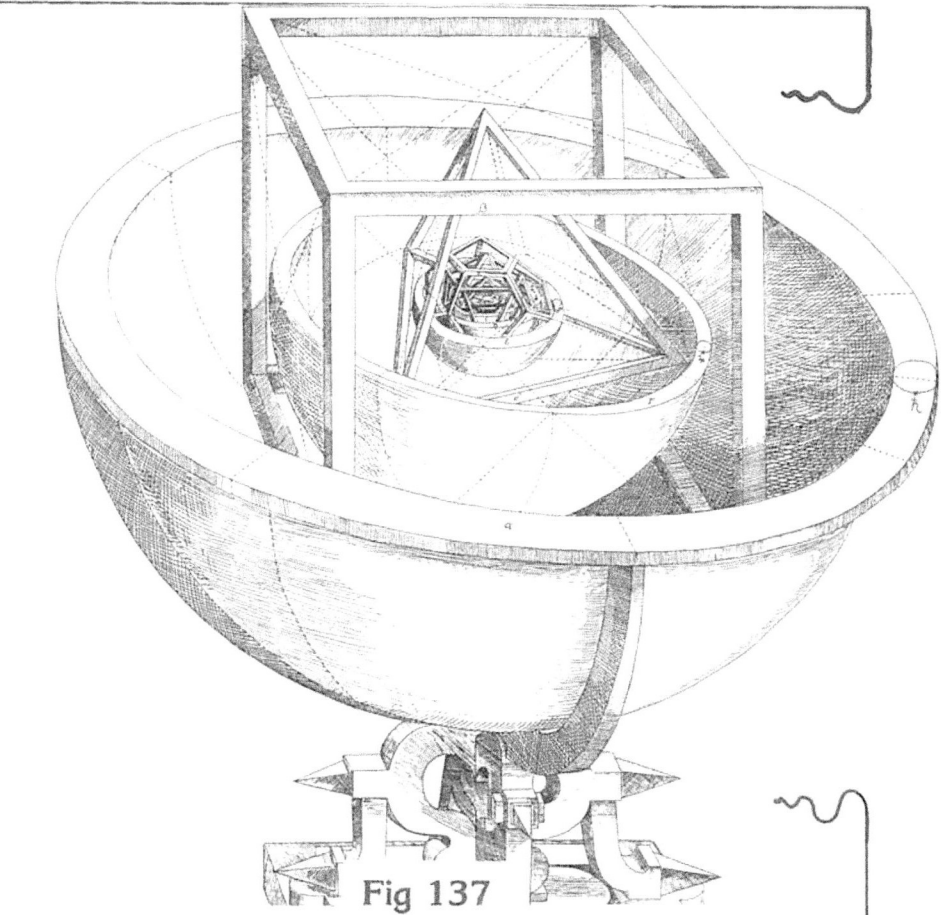

Fig 137
Kepler, in his 'Mysterium Cosmographicum' circa 1609, put forward a Sun-Centered planetary model explained as the nesting (like Chinese Dolls) of the Platonic Solid Symmetries and contradicted Copernicus' Earth-centered Solar System. He is often quoted for: "Geometry has two great treasures: one is the Theorem of Pythagoras; the other, the division of a line into extreme and mean ratio (ie: the PHI Ratio). The first we may compare to a measure of gold; the second we may name a precious jewel"

Fig 138
If someone asked you to define the "inclusive cascading harmonics" attributed to the Fibonacci Series (1, 1, 2, 3, 5, 8 etc) and the Golden Mean Phi Spiral, how it is not a number but a function, the best answer is none at all, but rather request they meditate upon the face of the ROSE FLOWER

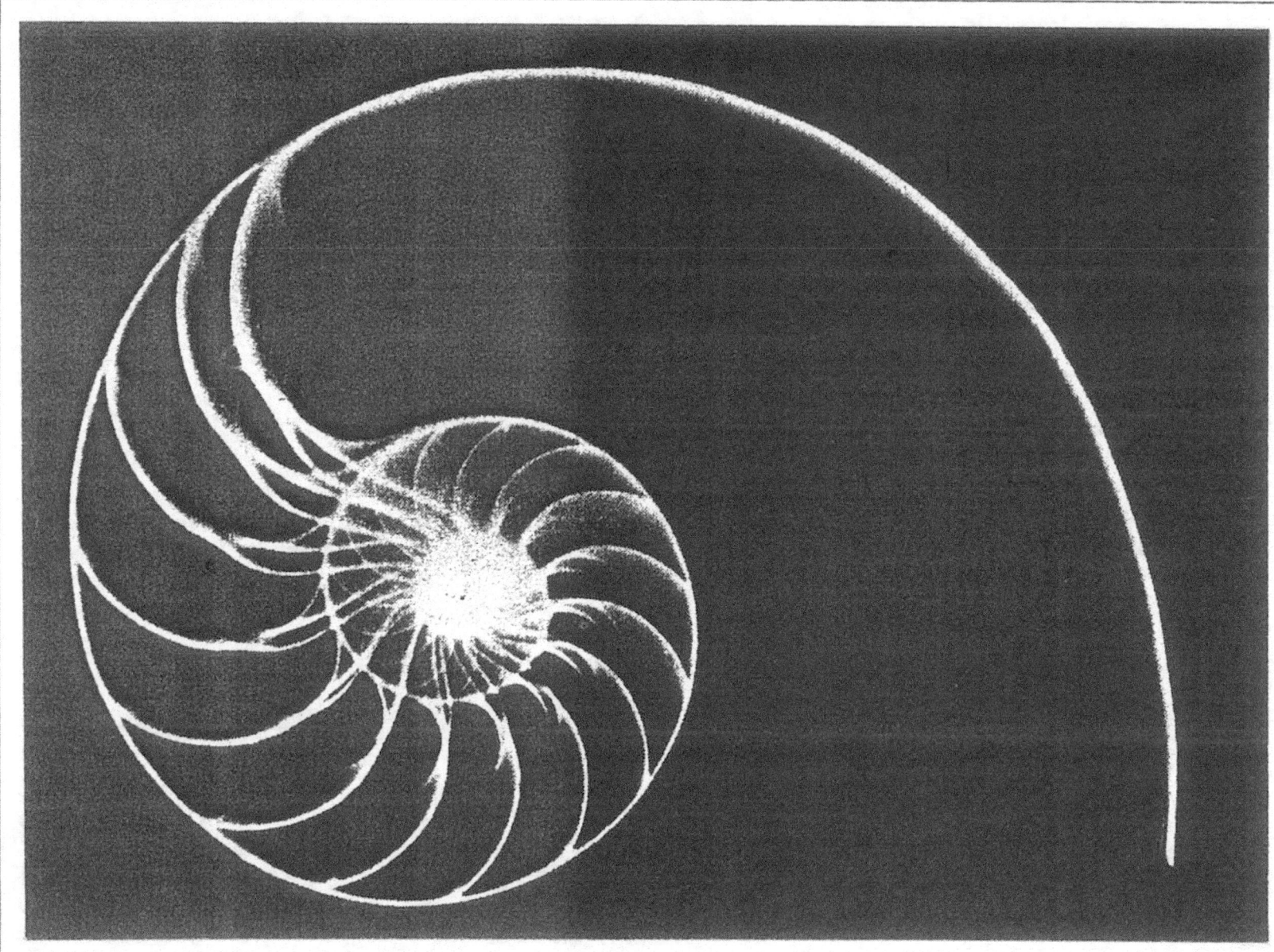

Fig 139
Radiograph of the well-known shell "Nautilus Pompilius". As it grows the volumes of the chambers increase in the phi ratio 1:1.618033 but the shape remains the same. Self-Similarity. The overall framework is defined as the Logarithmic Spiral.

ACKNOWLEDGEMENTS OF GRAPHICS USED AND NOTES.

LEGEND:
P = Page, PP = Pages, FIG = Figure, FIGS = Figures
RHS = Right Hand Side, LHS = Left Hand Side,
Bot. = Bottom, Cent. = Center Page, Lt = Latin,
ibid ='ibidem' Lt for 'in the same book, chapter, page'.
Each listed book has the Author's name referenced in full in the Bibliography.

P 24 FIG 10: Pentacle from 'A Beginner's Guide to Constructing the Universe'. See Bibliography.

P 32: The rose is taken from the front cover of "Here & Now" Magazine, printed freely in the Byron Bay area, Feb 2003 Issue 63. Dedicated to Freedom. See also Fig 138 on P163.

P 41 FIG 23: Research on Poisson's Ratio reproduced from www.silver.neep.wisc.edu/~lakes/sci87.html

P 44 FIG 26: from the 'Cube of Space'.

P 46 FIG 30: Alton of Sydney, studied with Drunvalo Melchizedek in America and created this 3-D rendition of his 'Flower of Life' Logo. See also P 72.

P 47 FIG 31: Gematria of the Pentacle from 'City of Revelation'.

P 48 FIGS 32a + 32b: from 'Rhythms of Vision'.

P 49 FIG 34: from http://mathworld.wolfram.com/Hypercube.html.

P 50: 'Shiva' from Moor's Hindu Pantheon.

P 52: 'No. 9 Dream' from http://www.cromp.com/tess/apc/page04.html.

P 58: Top RHS. 'Mitosis' from 'A Beginner's Guide To Constructing The Universe'.

PP 61–67 FIGS 42–48: from 'Alchemy Today'.

P 72: 3-Dimensional Flower of Life. See note for P 46.

P 75: Bot. RHS. Illusionary Dodecahedral shape modified from : 'A Beginner's Guide To Constructing The Universe'.

P 79 FIG 60: from Keith Critchlow, specific book unknown.

P 80 FIG 62: from 'Rhythms of Vision'.

P 83–85 FIGS 65–68a: Graphical Derivation of Phi from: http://www.vashti.net/mccinc/rgphigra.html 03/10/2002.

P 87 FIG 70: http://www.t0.or.at/c~scapc/c~mus14.html.

PP 88–89 FIGS 71–73: Rodin's Coil graphics from: http://www.innerx.net/personal/tsmith/SegalConf2.html.

P 90: Cent. from the book:'Yatri', picture called: 'Unknown Man: The Mystical Birth of a New Species' p208.

P 93 FIG 77: from 'The Inner Structure Of The I–Ching'.

P 100 FIG 83d: computer generated image by Thomas Banchoff and Charles Strauss.

P 100: 'Olivetti' symbol/calligraphy artist unknown.

PP 118–119 FIGS 96–97: from Buckminster Fuller.

P 124: from Alchemy Today.

P 126 FIG 106: Curve Stitching from 'The Nature of Mathematical Thought'.

P 127 FIG 108: photo courtesy of TRW Inc. 'STL' = Space Technology Laboratories, Inc. is a subsidiary of Thompson Ramo Woolridge Inc. Photo appeared in a book or chapter on 'Regular Polygons' p218.

P 132 FIG 114: 'The Cunning Caliph'.
P 134 FIG 117: from "Pascal's Triangle".
P 135 FIG 118: Spores from 'Introduction to Microfossils' redrawn from Hoffmeister, Staplin and Malloy, 1955.
P 135 FIG 119: Pollen from ibid, redrawn from Kuyl, Müller and Waterbolk.
P 136 FIG 120: 'Oracle of Delphi' from 'The Secret Teachings Of All Ages'.
P 139 FIG 122b: from 'Mathematics In The Making'.
P.141: Indian symbol for the sound of OM, the primal sound of Creation.
P 143 FIG 125a: from 'Curious And Interesting Geometry. A Dictionary'. Coaxial Circles: There are 2 sets of coaxial circles. One set consists of all the circles through 2 fixed points. Each circle of the second set is orthogonal or at 90° to every circle of the first set. Observe the enlarging concentric circles.
P 143 FIG 125b: ibid. Lemniscate of Bernoulli: (Lt: lemnicus = ribbon) by Jakob Bernoulli, 1694. Beginning with a rectangular hyperbola and drawing circles whose circles lie on the hyperbola and which go through the centre of the hyperbola creates the Lemniscate as an envelope. It is the inverse of the hyperbola, with respect to its centre.
P 154: LHS acronym (HOME) Home On Mother Earth by Patricia Diane Cota-Robles.
P 157 FIG 130g: 'Pillar Of Light' from 'Manual On Using The Light' by Joh Roger, 1976.
P 159-160 FIGS 131a+131b+132a: from 'Alchemy Today' which Chris Illert reproduced from 'Occult Chemistry'.
P 161 FIG 133: from 'The Web Of The Universe'.
P 162 FIG 135: Platonic Solid Duals from: www.mathworld.wolfram.com/DualPolyhedron.html.
P 162 FIG 136: from 'A Beginner's Guide To The Universe'.

P 163 FIG 138: see P 32.
P 164 FIG 139: photo courtesy of Kodak Limited, London. Reproduced from 'The Divine Proportion'.
P.166: Photo of Jain by Harmonie Downs in Byron Bay. 1997
P.180: Islamic Pattern + Art by Jain.

Photo of JAIN

~ BIBLIOGRAPHY ~

- **THE POWER OF LIMITS:** Proportional Harmonies in Nature, Art and Architecture. Gyorgy Doczi. 1981. Shambhala.
- **THE MATHEMATICS OF THE COSMIC MIND:** L. Gordon Plummer. 1970. The Theosophical Publishing House.
- **SACRED GEOMETRY:** Robert Lawlor. 1982. Thames & Hudson.
- **PLANET HEARTWORKS + ALPHABET OF THE HEART:** Dan Winter. 1988. Crystal Hill Farm.
- **ART FORMS IN NATURE:** Ernst. H. Haeckel. 1904. Dover 1974. "REPORT on the SCIENTIFIC RESULTS of the VOYAGE of H.M.S CHALLENGE" Volume 18, London. 1887.
- **THE DIVINE PROPORTION:** A Study In Mathematical Beauty. H.E. Huntley. 1970. Dover.
- **THE SECRET TEACHINGS OF ALL AGES:** Manly. P. Hall. 1988. The Philosophical Research Society.
- **THE ANCIENT SECRET OF THE FLOWER OF LIFE:** Volumes 1 and 2. Drunvalo Melchizedek. 2000. Clear Light Trust.
- **ALCHEMY TODAY:** Platonic Geometries in Nuclear Physics. Volume 1. Chris Illert. 1992. (Volume 2, published 1993).
- **OCCULT CHEMISTRY + CHAKRAS** (1927): Investigations By Clairvoyant Magnification Into The Structure Of The Atom. C. W. Leadbeater & A. Besant. First published early 1900s, then in 1951. The Theosophical Publishing House.
- **PYRAMID POWER + BEYOND PYRAMID POWER** (1975): G. Pat Flanagan, Ph. D. 1973. DeVorss & Co.
- **ISLAMIC PATTERNS + TIME STANDS STILL:** Keith Critchlow. 1976. Thames & Hudson.
- **BUCKMINSTER FULLER'S UNIVERSE:** An Appreciation. L. S. Sieden. 1989. Plenum.
- **THE CRYSTAL CONNECTION:** Randall & Vicki Baer. 1987. Harper & Row.
- **PYRAMID TRUTH GATEWAY UNIVERSE:** Reg Miller. 1997. Pyra Publishing.
- **THE TEMPLE OF MAN:** Volumes 1 & 2. R.A.S. de Lubicz. Inner Traditions.

BIBLIOGRAPHY cont...

- **THE SECRETS OF THE GREAT PYRAMID:** Peter Thompkins.
- **THE DIMENSIONS OF PARADISE + CITY OF REVELATION:** John Michell. Thames & Hudson and Abacus respectively. 1972
- **THE HIDDEN KNOWLEDGE OF GANN:** Les J. Clemens. 1991. +
- **GEOMETRY OF MARKETS:** Bryce T. Gilmore. Traders Press Inc. 1989.
- **THE STONES CRY OUT:** Bonnie Gaunt. 1991.
- **THE COSMIC OCTAVE:** Origin Of Harmony. Hans Cousto.
- **THE LANGUAGE OF PATTERN:** Keith Albarn, J. Smith, S. Steele, D. Walker. 1994. Thames & Hudson.
- **A BEGINNER'S GUIDE TO CONSTRUCTING THE UNIVERSE:** A Voyage From 1 to 10. Michael S. Schneider. 1995. HarperPerennial.
- **INITIATION:** Elizabeth Haich. 1965. Unwin. John Wiley & Sons.
- **VORTEX FLOW IN NATURE AND TECHNOLOGY:** Hans J. Lugt.
- **M.C.ESCHER KALEIDOCYCLES:** D. Schattschneider & W. Walker. 1977.
- **HARMONIC MATHEMATICS:** A Phi-Ratioed Universe as seen through Tone-Number Harmonics. William B Conner. 1982. Tesla Book Co.
- **TIMAEUS AND CRITIAS:** Plato. 1965. Penguin.
- **PATTERNS IN NATURE:** Peter S. Stevens. Penguin.
- **PASCAL'S TRIANGLE:** T. Green & C. Hamberg. + **VISUAL PATTERNS IN PASCAL'S. Δ:** Dale Seymour (Publications).
- **MORE MATHEMATICAL PUZZLES AND DIVERSIONS:** Martin Gardner. 1961. Pelican.
- **COURSE IN THE ART OF MEASUREMENT WITH COMPASSES AND RULER:** Albrecht Dürer.
- **THE PAINTER'S SECRET GEOMETRY:** C Bouleau.
- **THE WEB OF THE UNIVERSE:** E.L. Gardner. 1936. The Theosophical Publishing House.
- **MATHEMATICS IN THE MAKING:** L .Hogben. Galahad Books.

BIBLIOGRAPHY cont...

- **PRACTICAL APPLICATION OF DYNAMIC SYMMETRY:** The Law of Phyllotaxis. Jay Hambidge. 1920. Yale.
- **PATTERN AND DESIGN WITH DYNAMIC SYMMETRY:** Edward B. Edwards. 1967 Dover. New York.
- **THE MAN WHO LOVED ONLY NUMBERS:** Paul Hoffman. 1998. Fourth Estate Limited. (The Best Selling Story of PAUL ERDOS [pronounced: 'Air-Dish'] and the search for mathematical truth).
- **MADACHY'S MATHEMATICAL RECREATIONS:** Joseph S. Madachy. 1966. Dover.
- **MATHEMATICS AND THE IMAGINATION:** E Kasner and J. Newman 1949.
- **EXTRA-SENSORY PERCEPTION OF QUARKS:** Stephen M. Phillips. 1980. Theosophical Publishing House.
- **SEDONA: BEYOND THE VORTEX:** Richard Dannelley. 1995. Vortex Society.
- **THE UNIVERSAL ONE:** Volume 1. Walter Russell. 1926. The University of Science and Philosophy. 1974.
- **MAGIC SQUARES AND CUBES:** W. S. Andrews. 1917. Dover.
- **THE GAME OF SIGNATURES:** A Study of Gematria and the Numbers of God. Dr Lionel Joshua Hovey. 1993
- **MICROBIOLOGY:** (Introduction to Microfossils) Jacquelyn Black. 1993. 2nd edition.
- **THE PENGUIN DICTIONARY OF CURIOUS AND INTERESTING GEOMETRY:** David Wells. 1991.
- **THE PENGUIN DICTIONARY OF CURIOUS AND INTERESTING NUMBERS:** DAVID WELLS. 1986.
- **THE UNIVERSAL HISTORY OF NUMBERS:** Georges Ifrah. 2000.
- **THE CUBE OF SPACE:** Keven Townley. 1993. Archive Press.
- **FLATLAND:** Edwin A. Abbott. 1884.
- **RHYTHMS OF VISION:** Lawrence Blair.
- **THE CUNNING CALIPH:** Kenneth J. Kelsey. 1979.
- **MERLIN'S PUZZLER:** Volume 2. C. B. Townsend. Hammond Incorporated.
- **THE INNER STRUCTURE OF THE I-CHING:** Lama Anagarika Govinda.

INDEX

LEGEND:

aka = Also Known As.
Archi. = Archimedean Solid.
Cu.Nos. = Cubic Numbers.
D. = Dimension as in 3-Dimensional.
Dodeca. = Dodecahedron.
'e' = Exponential Function 2.718
f. = from.
"–" = Hyphen, eg 26–29 means from pages 26 to 29 inclusively.
lit. = literally
M.Sq. = Magic Square.
M.Sq.C. = Magic Square Constant.
Nos. = Numbers.
P.Sols. = Platonic Solids.
pics. = Pictures or Graphics.
pl. = plural.
Pyr. = Pyramid.
sing. = singular.
Sq. = Square.
Sym. = Symmetry or Symmetries.
Tet.Nos. = Tetrahedral Numbers.
Δ = Triangle.
Tet. = Tetrahedron or Tetrahedral.
VM = Vedic Mathematics.
www. = website address.

A

Abbott, Edwin A	101
Add and Multiply	22–23
Adonai	77
Adyarium 12	61–63, 65, 127–129
Alchemy	23, 61–67, 127–129
Algebra	53, 71, ('e')146–150
Al Jabr Khwarizmi	71
Allah	77
Allotropic	59, 61
Alpha and Omega	110
Alpha Particle	(see Adyarium 12)
Ammonium	60
Angels	57, 145
Angles	57, 60
Ammunition Store	138–139
Androgynous	72
Anti–Gravity	94
Anurupyena (VM),	(see Proportionately)
Anus	58
Apollo	136
Aryan	11
Archimedean Solids	37–38
(Archi. Dual)	57, 64, 69, 104
Argentite	38
Arithmetic Progression	83–85
Art by Jain	92, 96, 104, 131 152–158, 180
Art of Discovery	108
Atom	(Structure) 23, 27, 35–36
	60–67, 71, 80, 107, 127–129
Ashmolean Stones	69
Axis of Spin	112

B

Babylonians	105, (Sq.Pyr.Nos.) 138
Bacteria Population	('e') 147
Base 12 (see Number's Index)	53–54
Beryllium	63–64, 67
Besant, Annie	66, 127–129, 59–160
B-Field Torsion	94
Bibliography	167–169
Binary	88–97
Binomial Expansion	51
Bipolar Symmetry	89
Biune Trinity	19
Black Holes	87
Blair, Lawrence	80
Bragdon, Claude	25, 98
Brain	54, 106, 111–112, 115, 143, 145
Breathe	152-158
Brittanica Encyclopedia	112
Brooks, R	108–109
'By One More' (VM)	83, 103,120
	('e') 148
'By Mere Intuition' (VM)	122

C

('e') 148–149

'Calana Calanabhyam' (VM) see Calculus
Calcium 62, 127
Calendar 112
Calculators 7–9, 11, 26, 46
Calculus 8, 13, 55, ('e') 147–150
Canon Balls (count) 138–139, 140
Carbon 45
(CH_4 Carbon Tetrachloride) 60, 61–63
Cascade Between Frequencies 89
Cayce, Edgar 111
Centre of Mass 66–69
Chakras 114, 145, 152–158
Chinese Triangle (see Pascal's Δ) 133–134
Circles (as spheres) 144
Clairvoyant Scientists 61–69
Closest Packing 138–139
Coaxial Circles 143
Co-efficients 51–53, 55
(Pas. Δ + Pyr.) 134
Compound Interest ('e') 146–150
Compression 8, 24, 38, 55, 71
76, 88, 90, 127–129
Computer Chip (Silicon) 45
46, 79–80
Conclusion 145
Concrescence 71
Conjugate (Golden Ratio) 28

Consciousness 87, 111, (pure) 145
Consecutive Order 116, 120
(Δ Nos.) 121, (Cu. Nos.) 122, 151
(Tet. Nos.) 123, (Tet. + M.Sq.C) 141
(Factorials) 146, ('e') 149
Constant 112
Contact (the movie) 104
Conway, J.H 42, 99
Cosmic Energy (Earth-Heart) 152–158
Copernicus (Solar System) 163
Creation Geometries 57–59, 66
78–82
Critchlow, Keith 69
Cross (of Jesus) 79–80
Cross–Addition 16–17
Cross–Multiplication 29, 33, 54 72
Crystal 38, 45, (Beryllium) 67–69
80, 111, (Dodeca.) 131, 145, (Earth) 52
Cube 22–23, 25, 33, 35, 40–41
43–44, 49, 56–58, 63, 69, 73, 76
78–82, 103, 111, 119, 144
–Cubic Equation 51–53
–Cubic Numbers (see also Powers of 3)
51–53, 111, 114–119, 151
–Cubic Number Series 122
–Cubic Packing 118–119, 138
–Cubing Versus Tetrahedroning 119
–Double Cube 25, 33, 43, 45, 73, 81
–Hyper-Dimensional Cube 49
Cuboctahedron 37–39, 57, 64–65
69, 99, 103
Curie, Marie 63

Curriculum 23, 121
Curve-Stitching 126

D

'd' shell (Atom) 63
Dan Tien (Earth-Heart) 152
Dan Winter 8, 112
Diamond 45, 61, (playing cards) 79
118, 144
Diabolic Donut (M.Sqs.) 77
Diabolic HyperCube 76
Diatom 123
Differences (in the Phi Code) 91–92
(in Polygonal Nos.) 109
Differential 13–14, 29
Digital Sums (VM) 3, 8, 55
Dimensional Ladder 49, 82, 110, 112
Discriminant 13–14, 29
DNA 8, 23, 25, 48–49, 55, 73
91–92, 145, 153
Dodecahedron 8, 23–25, 37
39–40, 43, 45, 57, 63, 69, 98, 119
127–129, (memory in restructured water) 131
Dome (geodesic) 118
Don't Do That 145
Double Square 9, 11–12, 22, 43
Doubling Sequence 88–97
Duals (see Platonic Solids)
(Archimedean) 37
(Self-Dual Pentatope) 40
(Self-Dual) 41
Duplex 9, 13, 15–19

Dürer	47	
Dyad	56	

E

e=2.71828	(see Number's Index)	
	3, 82–85, 146–150	
Earth	(Planet) 106, (Grid) 145	
	(Energy) 152–153	
Earth-Heart Meditation	151–158	
Eco-sophical	108, 145	
Egypt(ian)	15, 111	
Economics	('e') 146	
Eight Original Cells	66, 76, 78–82	
Einstein	112	
Els (or Elders Race)	34	
Electrical Coil	90	
Electrical Energy	(in water) 131	
Electro-Hydro-Dynamics	95, 131	
Electrons	91	
Eleven-agonal	(Polygon) 109	
Ethane	60	
Etheric Templates	90	
Euclid	12, 26–27, 121	
Euler	146–150	
Exponential Function	(see 'e' or 2.718)	
Exponential Growth	83–85	
Exponents	(see Powers)	
Extreme and Mean Ratio		
	(see Phi Ratio), (Kepler) 163	

F

Face-Centered Cubic	138	
Factorial	146	
Feng Shui	33	
Fibonacci Numbers, Reduced	3	
	23–24, 36, 41, 43, 55, 57	
Fibonacci Sequence	3, 15, 23	
	32, 70, 85, 133–134	
Fifth Element	69	
Figurate Numbers		
	(also Polygonal Nos.) 42, 137–138	
Five-Cell	41	
Fixed Design	104, 112	
Flanagan, Patrick	131	
Flatland	101	
Flower Of Life	(3-Dimensional)	
	46, 72, 101	
Fontanelle	(Earth-Heart) 152	
Forgiveness	145	
Formulae	110	
	(Multi-Dim. Constants) 112, 120–121	
	(Gauss) 122, (M.Sq.C. + Tet.Nos) 130	
	(Sq.Pyr.Nos.) 141, ('e') 146–150	
Foster, Jodi	104	
Fourth Dimensional (4D.)	25, 37	
	40–42, 49, 57, 73, 75, 87–101, 104	
	111–112, 121, (Pas.Δ) 134	
–Fourth Dimensional Eye	122	
–Fourth Dimensional M.Cube of 3x3x3x3		
	111–112	
–Fourth Dimensional Δ Nos.	121	
Fractal	8, 34–35, 43–44, 75, 81	
	89, 113	
Franklin, Benjamin	96	
Free-Energy	94	
Frequencies	114, 145	
Fuller, Buckminster	38, 63–65	
	117–119, 126–129	

G

Galaxy / Galactic	71, 73, (nos.)104	
	107, 111, (Maths) 145, (Earth-Heart) 156	
Gap Circuit	88–97	
Gardner, Martin	77	
Gauss, Karl	122, ('e') 149	
Gematria	(see also Number's Index) 47	
Genesa	87, 99	
Genetic	56-58, (Masters) 153	
Geometric Progression	83–85	
Geometry	(C. Bragdon) 98, 108	
Gnosis	108	
God	30, (Self) 145	
Goddessence	153	
Gold	66	
Golden Rectangle	(see Phi Rect.)	
Golden Triangle	(see Phi Δ)	
Gratitude	157	
Graphite	44–45, 61	
Greater Maze	22–23	

H

Haeckel, E. H	161	
Halayudha's Tetrahedron	134	
Halayudha's Triangle	(see Pas. Δ)	
Hans Albrecht Bethe	62	
Harmonics	110–111	

Heart   46, 113, (Earth-Heart) 153
         (Anahata Chakra) 153–158
Helium   61–65
Hexa(gon)   23, 37, 43–49, 61, 81 89
         (faces in Magic Tetrahedron) 132
HecatonIcosahedroid   98
Heptagonal   109
Hitch-Hiker's Guide to the Galaxy   111
Hockey Stick Rule   (in Pas. Δ) 133–134
Holographic   25, 43, 106
         (hologram) 145
H.O.M.E.   (acronym) 154
Hydrogen   60–62
(Hydrogen Ions in restructured water)   131
Hypatia   121
HyperCube   (see Tesseract or 4D. Cube)
HyperDimensionalCube   49
HyperPolyHedra (Hyper Solids)   98
HyperSpheres   40, 42, 73
         (formula) 75, 87

I

'i' (Square Root of '–1')   ('e') 146–150
Icosahedron   23, 63, 69, 127–129
IcosaTetraHedroid   98
Illert, Chris   61–67, 127–129 159–160
Implosion   15, 20, 33, 35, 41, 49 55, 71
In-Cube-Ation   (Incubation) 45
Inclusive, Cascading Harmonics   163
         (in Rose Flower)

Indices   (see Powers)
Infinity (In-Phi-Net)   20, 23, 34–36 57, 71, 73, 75, ('e') 146–150
Inside Out   87
Iron   62
Isotopes   62, 159–160
Inter-Dimensional   35, 110
Intuition   ('e') 148–149

J

Jacob's Ladder   35, 111, 145
         (see also Dimensional Ladder)
Jaina   (Indian monks) 51, 90
Jesus Christ   47, 57, 79
Jitterbug   38–39, 64–65
Jupiter   106

K

Kali   50
Kepler, Johannes   163
Key   (ancient) 142
Khajuraho   (India) 77
Khem   (ancient name for Egypt) 66
King Solomon   9, (Rabbi's Golden Star Yantra at 0° + 90°) 67-69, 104
Kissing Number   42, 99

L

'L' Formations   117
         (Hockey Stick Rule) 133
Langham, Derald   56, 87, 99
Language of Light 107, 111–112, 145

Lattice   99
Law of the Squares   (Searl) 129
Lead   62, 66
Leadbeater, Charles   66, 127–129 159–160
Lemniscate   143
Lesser Maze   (Theosophy) 22–24
Linear Function   83–85
Logarithm   23, ('e') 147–148
Logarithmic Spiral   (Nautilus) 164
Logo   (Jain's M.Sq.) 113
Lost Sutra   (the 17th Sutra by Jain) 103
         ('e') 148
Love   145
Lucas Numbers   70

M

Macro(cosm)   27, 35, 43, 81
Magic Cubes   (Constants) 111, 145
Magic Cube of 3x3x3   110–112, 156
Magic Nucleus (or Nuclei)   61–62
Magic Number   63, 65
Magic Squares   (general) 35, (cubic nesting) 44, 104–114, 145
Magic Square Art   66
M. Sq. Pandiagonal or Nasik   77
Magic Square Constants   3, 63 76–77, 102–114, 106–109, 123 129–130, 137, 142–143, 145
–M. Sq. of 3x3 (Lo-Shu)   28, 50 65–69, 93, 104–105, (Logo) 113
         (Constant = 0) 144, 155

–M. Sq. of 4x4	113, 132
(Magic HyperCube/Tesseract)	76–78
–M. Sq. of 5x5	106
–M. Sq. of 6x6	106, 129
–M. Sq. of 7x7	106
–M. Sq. of 8x8	96, 106
–M. Sq. of 9x9	105–106
–M. Sq. of 10x10	106
–M. Sq. of 16x16	96
Magic Tesseract	25
Magic Tetrahedron	132–133
Magnetic Fields	73, 86–97, 94
Mandala	49, 124, (Sri Yantra) 144
Manjushri	93
Many Dimensions	112, 144
Mars (M. Sq. of 5x5)	106
Masons	80
Mathematical Monk	108
Meditation (Earth-Heart)	151–158
Memory	23, 25, 56–58, 90, 107
	111, (P.Sol. Symmetry) 131
	(Banks) 145, (Earth) 153
Merkabah (Light Vehicle)	75, 81, 155
Meru Foundation (Stan Tenen)	144
Meru Prastera (aka Pascal's Tri.)	90
Methane	60
Michell, John	47
Micro(cosm)	27, 35, 43, 81
Micro-Fossils	132, 135
Micro-Psi (Occult Science)	63, 159–160
Midpoints	71
Mirror Image	92, 96–97

Mitosis	58, 99
Monad	56
Moon	105–106
Mouse Egg	56
Mouth	58
Mudra	152, 155
Music	97
Multi-Dimensional Constants	112
Mysterium Cosmographicum (Kepler)	163

N

Natural Base (Logarithms 'e')	146–150
Natural Square	66–69, 104, 106
Nasik (M.Sq.)	77
Naudin, J.L.	86–97, (www.) 94
Nautilus Pompilius (Phi Spiral Shell)	164
Navigational Tables ('e')	147
Negative Numbers	8, 13–14, 22
28–29, 71, 85, (in M.Sq.C=0)	114
Negative Phi or '–Φ' (see Number's Index)	
	= –1.618033
Negative Reciprocal of Phi '–1/Φ'	
(see Number's Index)	= –.618033
Nesting (M.Sq.3 in M.Sq.4)	113, 118
(M.Sq.C.) 124, (P.Sol.Sym.)	127–129
(toroidal)	157
Nets (of Tet.+Icosa.+CubOcta.)	65
Neophyte (Earth-Heart)	154–155
Neurophone	131
Neuro-Transmitters	111
Nine-Point Circle	97

Nonagon(al)	97, 109
Now	145, 153
Nowhere = Now–Here	153
Numerical Nomad	108
Numerology	90

O

Occult Chemists	61–67, 127–129
	159–160
Octagonal	109
Octahedron	23, 69, 78–79, 144
Odd Numbers	115–116, 137
Olivetti	100
Oracle of Delphi	136
Oranges (pyr. packing)	138–139
Orion Constellation	111
Over-Unity	95
Ovum	57, 62–65
Oxygen	60, 63–65, 127–129

P

'p' shell	63–67
Painter's Secret Geometry (Phi based)	47
Pairs (Lo-Shu) 93, (Gauss)	122
Pairs: (12 Pairs of 9) (see also	
Number's Index for '9' and '12')	
3, 24–25, 28, 36–38, 40, 53–54, 57	
61, 69, 78, 81, 86, (Sq.Pyr.Nos.)	142
Palindromic	38
PanDiagonal Magic Square	77
Parabola	83–85

Pascal's Triangle	51, 90, 124	
132–134, (Pas.Pyr. in 3-D.+4-D.) 134		
(Sq.Pyr.Nos.) 140		
Path of Least Resistance	118	
Pattern Hunter	108, 122	
Pattern Recognition	3, 16, 108, 143	
Peanuts (cartoon)	150	
Pentacle ☆ 24, (4-D.) 25, 39–40, 43		
45–49, (of Rabbi Solomon the King) 69		
Pentagon	19, 37, 39, 40–41	
48–49, 98, 144		
Pentatopes	37, 40–41	
Perfection (M.Sq.C=0)	114	
Perineum (Earth-Heart)	152	
Periodicity	3	
Periodic Table	64–66, 159–160	
Permutations (of M.Sq.4)	132	
Phi (General)	146	
–Phi Code	3, 36–43, 61, 76, 81	
(Differences) 91–92,		
–Phi, Derivation of	83, 85	
–Phi-Loso-Phi	53	
–Phi Poem	34, 53–54	
–Phi Ratio	113, (Kepler) 163	
–Phi Reciprocal (see Nos. Index) =.618		
12, 19–21, 23, 28, 33, 40		
–Phi Rectangle	48, 82–83	
–Phi Spiral	8, 34, 48, 55, 107, 145	
151, 159–164		
–Phi Triangle	66, (in M.Sq.3) 113	
–Phi Vesica	73	

Phosphorus	59
Pi (See Number's Index), ('e')	146–150
Pineal Gland (3rd Eye)	9, 46
Planetary Squares	105
Plato	69
Platonic Fire	62
Platonic Solids (General)	25, 38, 63
65, 104, 112, 118–119	
124, (water) 131	
Platonic Solid Duals	28, 78–79
(diagrams) 159–164	
Polygons	37–38, 41, 48–49, 162
Polygonal Number (aka Figurate Nos.)	
108–109, 138	
Polyhedron	24, 41
(HyperPolyhedra) 98, 160–163,	
(Duals) 162, (Compounds) 162	
Pollen	132, 135
Powers (aka Exponents, Indices)	
3, 24, 30, 70	
82–85, 88–92, (Similar) 146, 147–150	
–Powers of 2 (aka The Squared Nos.)	
88–89, 104, 115–119, 133–134	
–Powers of 3 (aka The Cubic Nos.)	
111, 114–119	
–Powers of 11	90, 133–134
–Powers of Phi (aka Lucas Nos.)	70
Pranic Tube	152
Predict Sequences	108, 115, 121–122
Prime Numbers	104, ('e') 148
Probability, Theory of ('e')	147

Problem Solving	104
Prophetess	136
Proportionately (VM Sutra)	29, 41
43, 50–52	
Proteins	45–46, 48
Protons	61–67
Psycho-Active (Mandalas)	124, 143
Pyramid	111, (of addition) 115–116
(Pas.Pyr.3.D.) 134, (Sri Yantra) 144	
Pyramid Numbers (See Square Pyramid Numbers)	
Pythagoras	19, 27, 120–121
136, (Kepler) 163	

Q

Quadratics (Equations + Roots)	8, 9
11, 13–14, 19–21, 29–34, 54–55	
72–74, 85, ('e') 146–150	
–Quadratic Formula	10, 13
–Quadratics (Dual Roots of Phi)	30–33
Quantum Physics	66–67, 127–129
Quarks	62
Quartz (Silicon Dioxide)	60

R

Radioactive	63, (Decay)+('e') 147
Radiolaria (Protozoa, single-celled)	161
Rain Man (the movie)	122
Reciprocals	15, 20, 33, 70
Reciprocation (Polyhedron Duals)	162
Recursion	3, 23–24, 36, 55
57, 86, 88–97, 114	

Restaurant at the End of the Universe 111
Reverse Order (Gauss) 122
Rhombus 46, 118–119
Rodin, Marko 86–97
Roots (graphic) 27, 34, 81
Rose 32, 163

Russell, Walter 82
Rosser + Walker 77
Row-By-Row Analysis 67–69

S

's'-Shell 62–67, 127–129
Sacred Geometry 144–145, 162
Sacred Science Institute 35
Samuccaya (VM Sutra) 55, 88–91
Satellite 127
Saturn 105
Scale Invariant 43, 55, 71
Schodinger 128
Sea Urchin 56, 87
Searl, J.R 129
Self-Similarity 32, 43, 113
   (Nautilus Shell) 165
Shakti 50
Shape 25, 35, 38, 43–44, 47–48
   107, 110–111, 115, 122
Shareable 108
Shiva 50
Silicon (chip) 45, 59
   (Dioxide) 60, 79–80
Silver Ratio (Phi) 28

Silver Sulphide 38
Sino-Tibetan 93
SkyHeart (Meditation) 153, 152–158
Slinky 73
Sloane, N.J.A 42, 99
Soap Bubbles 59, 75
Soddy 62

Solar System (Kepler+Copernicus) 163
Solid Tessellation (Tetrahedral) 126
Solomon (see King)
Spacecraft (see also Merkabah)
   89, 94, 129
Sperm 56–57, 87
Sphere 36, 41, 56, 59, 73,
   (formula) 75, 79, 81, 87
   99–100, 103, 144
Sphere-Packing (see also Closest
   -Packed Symmetry), 3, 41, 43, 57–61
   63, 76, 99, 102–103, 118
   (of M.Sq.C.) 124, 138
Spin 22, 35, 82, 112, 114, 145
Spores 132, 135
Squares 10, 45–47, 38–40, 42
   49, 57, 144
–Double Square (or 2x Unity's Square)
   9, 11, 29, 31, 43, 50, 81
Squared Numbers (see Powers of 2)
Square Pyramid Nos. 104, 137–143
Square Roots 15, 18–19, 81
   (see also √5 in Number's Index)
Sri Yantra (aka Shri Chakra) 144

Stack-Base Number 128
   (see also Tetrahedron Stack Number)
Stack Number 63
Star of David ✡ (aka Star of Solomon)
   39, (elongated) 67–69
   (f.M.Sq.3) 104, 144, 155
Star Tetrahedron 23, 37, 45, 57, 63
   66–69, 80, 111, 144, 155

Stella Octangula (aka Star Tetrahedron)
ST ND RD 12
Sunflower Map (Rodin's Coil) 88–89
Sunlight 62, 106, 157
Symmetry 92, 96–97
   (in dodecahedrally restructured water) 131
Swaztika (in 3D.) 43

T

Tantra 136
Technology 45–46, 71
Tenen, Stan (3D. Sri Yantra) 144
Tessellations 43
Tesseract (aka 4D. Cube or HyperCube)
   25, 35, 37, 40, 43, 49, 57, 73
   (Diabolic HyperCube) 76, 77–78
   111–112
Tetrahedral Consciousness 107
   (Art by Jain) 131
Tetrahedral Nos. (aka Tetrahedronal Nos.)
   102–104, 106–107, 118, 120
   123–136, (Triples of) 130, 138
   141–142

Tetrahedral Tripod 136
Tetrahedron 3, (4D.) 37, 41, 43, 45 56–59, 76, 78–80, 102–103, 107 111, (Cube Vs Tetra) 119, (Stack Nos.) 126, (Satellite) 127, 128–129, (Micro-Fossils) 135, (God) 145, 59–163
Tetrahedron, 4th Dimensional (see Pentatope)
Tetrahedron Stack Number 63–64 127–129
Tetraktys (Pythagorean) 120
Theon 121
Theosophy 22 (Theosophical Society) 159–160
Third Eye (see Pineal Gland)
Tibetan 93
Timaeus (book by Plato on P.Sols.) 69
Time Codes 112
Time Travel 22, 34, 44, 110–111
Torus (aka Toroidal Domain), 27–28 34, 56, 58, 73, (nesting Tori) 74 76–78, 86–97, 104, 145, 157
Trefoil (Trifolium: Lt '3-Leafed) 79
Triakis Tetrahedron 37–38, 57
Triangle (equiangular) 10 (right-angled) 26, 38–39, 42–43 49, 64, (Sri Yantra of 9 Interlacing) 144
Triangling Versus Squaring 118
Triangular Numbers 104, 108–109 115–116, 120–122, 128 133–134, 137

Trigrams (f. Chinese I-Ching) 93
Trilete Sutures (spores) 135
Trinity 19, 29–30, (Trinary) 90
Trinomial (in Pas. Pyramid) 134
TriVision 29, 54, 72
Truncated Tetrahedron 37–38, 57
Truth 153
Twelve-Agons (Polygon) 108

U

Union of Hex and Pent 47–48
Unlimited Series ('e' with a Limit!) 150
Unity 23, 34, 54, 71, 154
–Internal Division of Unity's Square ($x^2+x-1=0$) 26–31, 73–74
–External Division of Unity's Square ($x^2-x-1=0$) 26–31, 73–74
Unity Consciousness 20
Unity Trivided (see also Trivision) 19, 29, 54, 72

V

Valence 88, 129
Vastu Shastra (see Feng Shui)
Vedic Mathematics 8, 15, 29 46 50–55, (Poem) 53–54, 66, 83, 88 (Jain's 17th Sutra) 103, 111–112, 120 (Jain's 17th Sutra) 122, ('e')124, ('e') 149
Vedic Seers 112
Vehicle of Light (aka Merkabah) 110, 155
Venus 106

Vertex [(sing.) Vertices (pl.)] 40, 98 119, 127–129, 162
Vertically + Crosswise 15–17, 55, 106
Vesica Pisces (lit. 'Vessel of the Fish') 9, 46–47, (Phi Vesica) 73–75, 162
Visualise 145
Volumes (P.Sols.) 162, (nautilius) 164
Vortex 86–97, (Maths.) 89, (pics) 94–95

W

War 145
Water 45–46, (on Two Legs) 111 (restructured water) 131
Watson and Crick 48
Wormholes 87

X

'X' Factor (see also 'Vertically + Crosswise') 71, 106

Y

Yantram [(sing.) Yantra (pl.)] 113

Z

Zero 15, 22, (M.Sq.C.=0) 114
Zero Point Technologies 114
Zodiac 93
Zygote 56–57, 99

THE NUMBER'S INDEX
OR THE HARMONIC STAIRWAY

The **HARMONIC STAIRWAY** is a vast unpublished Dictionary of Numbers, in the author's private Library, that was filed as a result of extensive research on Sacred Geometry. This 24 year compilation is a vast accumulation of all numerical references covering such subjects like:
- all the whole numbers up to and beyond the billions.
- all fractions of interest.
- all decimals.
- all negative numbers.
- all square roots, cube roots etc and all powers to the nth degree.
- all figurate number sequences eg the Pentagonal No. Sequence.
- all complex, imaginary and unending, infinite numbers like the famous: Π, Φ, e, i etc.
- all Hebrew and Greek gematria.
- all metaphysical sources like Magic Square and Magic Cube Constants which were the original inspirational reason for indexing such data.
- all multi-cultural sources like Indian, Egyptian, Pythagorean, astronomical and astrophysical.
- all scientific frequencies and Hertz (cycles per second) of planets, human organs, musical notes and more.

I always felt there was a need for such a dictionary of numbers, not knowing that already 2 distinguished sources have been published:
- D.Wells: "A Dictionary of Curious and Interesting Numbers + (A Dictionary of Curious and Interesting Geometry).
- Sloane's: "On-Line Encyclopedia of Integer Sequences".

If there is any particular fascination you may have as part of an advanced research for any particular number, I am able to pull out some or much foldered information for that specific number, vibration or frequency.

The Dictionary is also richly illustrated. Please contact the Author if you can assist in creating this data into a much needed CD Rom.

number's index

−1.6180339887498948482045868	(Minus/Negative Phi aka Φ) 19–21, 23, 28, 30, 43, 55
−1	20, 22
√−1	("i" or Sq. Root of Minus 1) 146–150
−.6180339887498948482045868	(Negative Reciprocal of Phi: −1/Φ) 29–30
−.38196601125020	($-1/\Phi^2$) 85
0 (ZERO)	15, 22, 114
.000101020305081 3	6
.38196601125020	($-1/\Phi^2$) 40, 71, 85
.5	19, 54, 70
.61803398874989484820	(1/Φ) 12, 19–23, 28, 33, 40, 54–55, 70–72
.888	47
1	34, 47, 70–71, 83–85
1.414213562	(or √2) 10
1.480	47
1.6180339887498948482045=Phi	
1.732050808	(or √3) 10
2.236067978	(or √5) 9–12, 14, 18–19, 22, 24–25, 28–29, 33, 43, 50, 55
2.368	47
2.618033988	(Φ^2) 70
2.7182818284590452353602 8	('e') 3, 102, 146–148
4	57–62, 88, 103, 107, 119, 123–134
4.236060	(Φ^3) 70

5	19, 21, 24, 41, 44–49, 137–142
6	38, 42–49, 64, 79, 81, 99, 128
6.854085	(Φ^4) 70
7	81, 87, 105, 112, 117
8	57, 66, 70, 73, 79, 88, 93, 103, 119, 122, (reclined) 154, 156
9	24–25, 36, 38–40, 50, 55, 81, 88–89, 92–93, 96, (9-Point Circle) 97, 118, (Sri Yantra) 144, 151, 155
10	28, 40–41, (decimal) 90, 93, 107, 123, 128, ('e') 147–148
11.09013	(Φ^5) 70
12	24–25, 37–42, 53–57, 61–69, 76–79, 81, 86–97, 99, 103, 119
13	25, 37, 57, 70, 81, 111
14	132, 137–142, (Earth-Heart) 156–157
15	28, 93, 102, 106–107, 109, 112, 121–123, 128, 130, 132, 137, 142–143
16	57, 64–65, 76–77
17	(17th Sutra) 103, (17th Sutra) 122
17.94420	(Φ^6) 70
20	64, 107, 119, 123–134
21	70, 117, 128
24	3, 36–38, 40–43, 55, 57, 76–77, 86, 89–100, 127–129
27	119
28	117, 128
30	137–142

32	58, 88
34	70, 76–78, 102, 107, 109, 112, 123, 130, 132, 137, 142–143
35	121, 123–34, (tetrahedral) 133
36	121
40	99
42	111–112
55	70, 137–142
56	(Tetrahedral) 123–124
60	45
64	58, 88, 93, 119, 122
65	102, 107, 109, 123, 130, 137, 142–143
70	121
72	99
84	(Tetrahedral) 123–124
89	70
90	34–35, 45, 68–69, 82
91	137–142
108	8, 24, 37, 55
109.5°	60
111	102, 107, 109, 123, 137, 128, 130, 142–143
120	98, 123–124, 130
123	112
130	112
140	137–142
144	70, 108

number's index

165	123
175	102, 107, 109, 123, 130, 137, 142–143
204	137–142
220	(Tetradhedral) 123–134
260	102, 107, 109, 123, 130, 137, 142–143
286	(Tetradhedral) 123–134
366	112
369	107, 109, 130, 137, 142–143
384	76–77
505	107, 109, 130, 137, 142–143
514	112
600	98
666	121
671	106, 130
720	98
880	76, (types/permutations of M.Sq.4) 132
888	47
1,200	98
1,480	47
9,899	6

by Jain. 1980.

Even a stone has a Consciousness!

Do you not think that the stone upon which a Master sculpts and chisels is not in rapture. The stone expresses deep gratitude in the fact and act of it being chosen to become a famous statue that is appreciated for a thousand years.

Carved in a Divine Proportion, so well modelled of the human form, it crys and sings in utter stillness.

It is Alive.

(Jain, 2003)

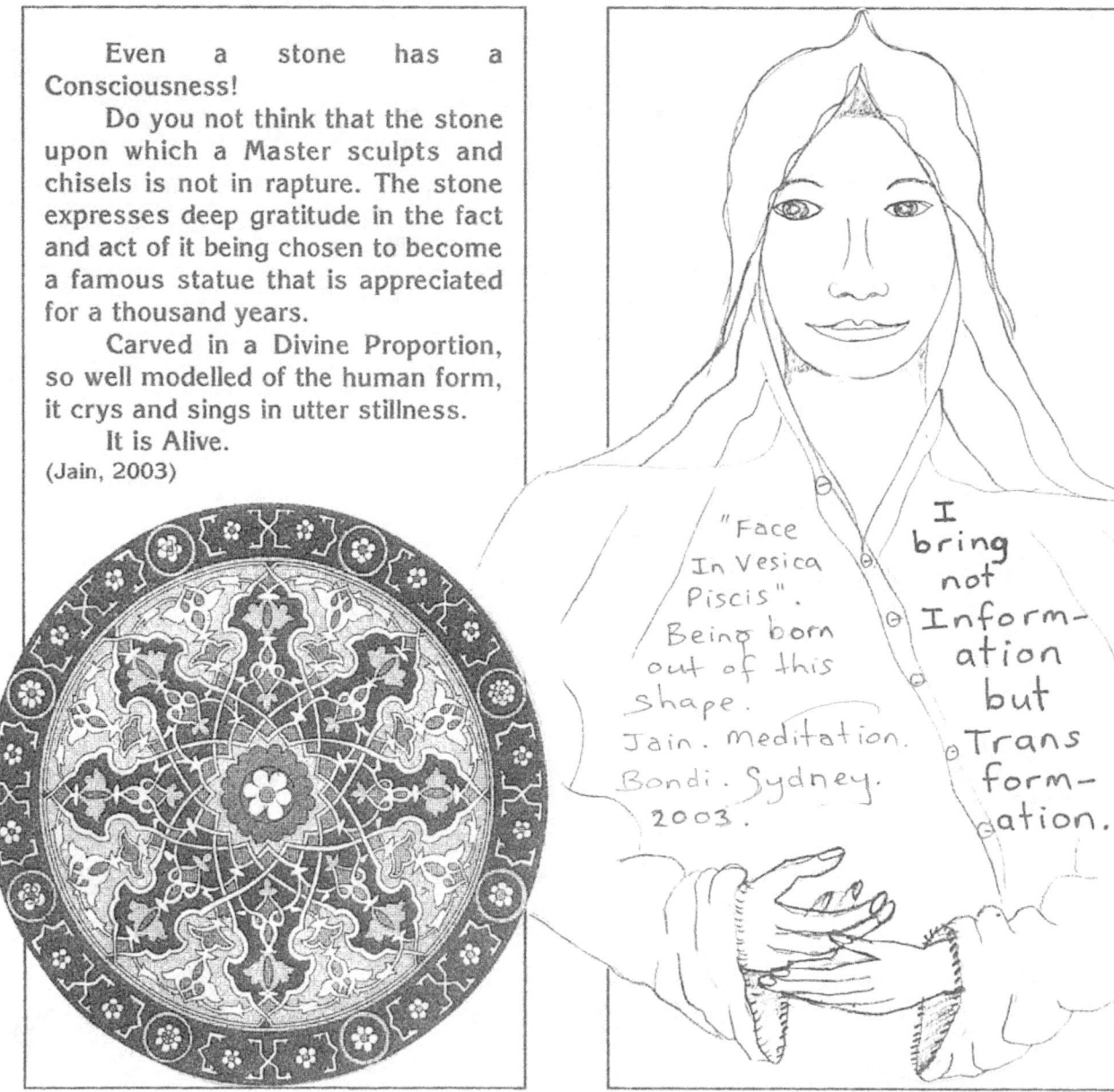

"Face In Vesica Piscis". Being born out of this shape. Jain. meditation. Bondi. Sydney. 2003.

I bring not Information but Transformation.

A DESCRIPTION OF JAIN'S BOOKS AND VIDEO

LEGEND:

'ISBN' = International Standard Book Number'
'pub' = published, 'pp' = pages, 'gms' = grams,
'RRP' = Recommended Retail Price, 'cov' = colour of cover,
'A4' = size, 'sb' = spiral bound, 'hw' = hand-written in calligraphy,
'ed' = edition, M.Sqs. = Magic Squares.

– 1 –
THE BOOK OF MAGIC SQUARES
Volume 1
formerly known as:
THE BOOK OF HARMONY SQUARES

ISBN: 0-9594180-1-6, RRP = $40, pub 1990 and 1995
ed 2, pp 164, gms 600, A4, sb, cov Magenta, hw.

This is a draw-your-own mandala colouring in book for children and adults. It consists of Magic Squares of 3, 4, 5, 6 and 7 selected from European, Essene, Tibetan, and Indian cultures. The Magic Square of 3 x 3, which is the centre of the Tibetan Calendar, when drawn and tessellated produces the atomic structure of Diamond Lattice, confirmed by the clairvoyant chemists Charles Leadbeater and Annie Besant 100 years ago. Magic Square Art forms create whole-brain learning as the translation of Number (Left Brain / logical) into Atomic Art (Right Brain / Pattern Recognition) allowing the student to absorb this ancient knowledge subliminally.

This book, was formerly known as:
THE BOOK OF HARMONY SQUARES, ISBN: 0-9594180-1-6, ed 1, pub 1990, size A3, perfect bound, print run 1000 copies sold). At that time, I gave myself the poetic and intellectual licence to consciously change the ancient name of the word "Magic" Square to "Harmony" Square, due to an unusual and fearful response from many Christian based educational institutions and Queensland Curriculum makers who objected to the word "Magic" (and the subsequent fusion of Western Cross/Eastern Swaztika-like patterns duly formed), despite "Magic" Squares being part of the mathematical curriculum studies. After a lot of "internal dialogue" about this issue, I decided to reclaim its former title of "Magic" instead of "Harmony" when the first 1000 A3-sized books sold, and created the second edition A4-sized with the current title.

– 2 –
THE BOOK OF MAGIC SQUARES
Volume 2

ISBN: 0-9594180-6-7, RRP = $40, pub 1996.
ed 1, pp 160, gms 600, size A4, sb, cov Light Blue, hw.

DESCRIPTION:

This book continues from where Book 1 finishes and consists of the Magic Squares of 8, 9 and 10. There is a whole chapter (Chap. 8) on and a dedication to Benjamin Franklin's Magic Square of 8 x 8. Of the 1,000 and more patterns I have constructed, I believe this to be the most harmonically and mathematically exceptional of all. For this reason, there is a window decal based upon this M. Sq. of 8 x 8 (see product 10B).

Chapter 9 includes a detailed account of the Indian Yantras or Power Diagrams symbolised by the 9 interlaced triangles of the Shri Yantra or Shri Chakra. The 9 triangles represent the 9 aspects of the divine feminine, like Shakti, Durga, Kali etc. In Sacred Geometry, it is helpful to visualise the 9 triangles as 9 inter-penetrating tetrahedrons.

- 3 -
THE BOOK OF MAGIC SQUARES
Volume 3

ISBN: 0-9594180-5-9   RRP = $40,   pub 2000
ed 1,   pp 174,   gms 600,   A4,   sb,   cov Purple,   hw.

DESCRIPTION:

This book continues where Book 2 finishes and consists of the Magic Squares of 11, 12 and 16. In Chapter 11, one particular M. Sq. Pattern or Yantra corresponds or looks similar to the Earth Grid as delineated by Bruce Cathy in his books Harmonic 33 and Harmonic 695 etc.

There is also a fascinating chapter on the Vedic Square known in this book as "The Visual Multiplication Table and its Reduction Grid". By mere application of a Vedic Sutra where we add the double digits of the Times Table and reduce them to single digits from 1 to 9, a whole universe of Atomic Structures are revealed.

Jain has spent 24 years teaching this in schools around Australia as part of his traveling Maths Enrichment Show called "Mathemagics". There is a lengthy and informative Glossary in the end section of this book not found in any of the other books.

- 4 -
JOIN-THE-DOTS
subtitled: The Art Of Magic Squares

ISBN: 0-9594180-4-0   RRP = $25,   pub 1999
pp 56,   gms 230,   size 210x210mm,   sb or stapled,
cov Violet with a Gold Star.
This small book is the first and only fully computerised book.

DESCRIPTION:

This smaller book was written specifically for the schools and is a summary of 'The Book Of Magic Squares', Vol 1. It has only the M. Sqs. of 3, 4, 5, and 6. It does not contain all the esoteric inclusions and original pen drawings as found in Volumes 1, 2 and 3 but represents the whole gamut of Magic Square Artforms in a simple and eloquent way. eg, when the simple M. Sq. of 3x3 is tiled 25 times the student has accomplished a beautiful Escher-like creation but is not informed that it is the Atomic Structure of Diamond Lattice (according to Leadbeater's Clairvoyant Chemistry)!

- 5 -
THE MAGIC OF NINE
IN VEDIC MATHEMATICS

ISBN: 0-9594180-7-5,   RRP = $40   pub 2001,
pp 124,   gms 390,   A4,   sb,   cov Yellow,

DESCRIPTION:

A highly intelligent and sophisticated system of Intuitive or Clairvoyant Mathematics. This book is inspired by Shankacharaya of Puri a.k.a. Bharati Krsna Tirthaji's (1884-1960) now famous book 'Vedic Mathematics'. This book has spawned a 1,000 websites! Shankacharaya was invited to America by Paramahansa Yogananda on a world tour spreading the ancient message of Vedic Mathe matics.

Shankacharaya's chief and current disciple is Maharishi Mahesh Yogi of Transcendental Meditation fame. Maharishi's spiritual school is churning out child geniuses like the 'Rain Man'. The Art of Mental Calculation is an important way of exercising the Brain's 'Mental Muscle'. In contrast to the left-brain-logical-male method of rote-learning we learnt at school, this system uses the right-brain-female style of Pattern Recognition. Its purpose is to make it known that we are far more superior than any calculator. Jain believes that if we persist to use calculators for the next 20 years, the global brain will atrophy. This book has a list of the 16 Vedic Sutras and explains some of them.

This book evolved as a series of lecture notes. Over the year whilst I was first demonstrating the definition of Vedic Mathematics as a Mental and One-Line Arithmetical System of 16 Sutras, people in the audience were too busy scribbling the examples down and therefore not paying full attention, so I decided to put most of the lecture examples and extra notes into this book.

- 6 -
THE VEDIC MATHEMATICS CURRICULUM FOR THE GLOBAL SCHOOL
PART 1: DIGITAL SUMS (of a 4 Part Series)

ISBN:0-9594180-8-3,　　RRP = $40,　　pub 2001,
pp 142,　gms 430,　A4,　sb,　cov Orange,　hw,
DESCRIPTION:

This book is everything possible on manipulating multi digits to single digits. This is the essence of 1 of 16 Vedic Sutras that means 'Digital Sums' or Digital Compression. Physicists know that if there is a problem to be solved, that compression is the answer eg if a computer has too much data and it needs to be stored economically or sent efficiently, it can be solved by digital compression.

A numerical example is, to multiply 33 by 11. We add the digits of 33 which is 3 + 3 which is 6 and insert this between the two threes, giving the instantaneous answer of 363.

Thus, 25 x 11 = 2　(2 + 5)　5 = 275. Its that easy.

There is a whole chapter on Multiplication by 11 and set out with questions and answers in the back. This is the essence of Vedic Mathematics, that it can be done mentally, via pattern recognition, or it can be done by one single line of intelligent arithmetic.

Chapter 1 is about "How Do I Start Teaching My Child The First Principles Of Vedic Meathematics?"

In Chapter 3 there is an indepth study of Universal Number Series like the Triangular Numbers and Pascal's Triangle which was really Halayudha's Triangle (a Vedic Jaina Mathematician from the C3rd BC). Of great interest is the Digital Compression of the Divine Proportion based on the Fibonacci Series: 0, 1, 1, 2, 3, 5, 8, 13, 21, 34, 55, 89, 144 etc and reducing them to single digits to discover eternal and internal patterning dealt with in full in: "The Book Of Phi", Volumes 1 and 2.

This is a great workbook for children 12 years and older. It will vastly improve the student's memory skills, increase their mental agility and boost self-confidence. Of the 4 Parts in this Series, Book 2 (to be published in Oct 2004) will be on all aspects of Multiplication; Book 3 will be on all aspects of Division; and Book 4 will be on Fractions, Calculus, Square Roots etc.

- 7 -
VIDEO: VEDIC MATHEMATICS FOR THE NEW MILLENNIUM

ISBN : 0-9594180-9-1　Recorded live 2002, 135 minutes,　430 gms, RP = $40　Format: VHS and NTSC available for the American market.
DESCRIPTION:

Filmed at the Cape Byron Steiner School, this video will give you a broad overview on how the 16 Sutras can solve every known mathematical problem. You will never multiply or divide again the way you were taught at school, once you adopt these basic principles. Learn why NASA are secretly using this system of Vedic Mathematics in the field of advanced robotics aka artificial intelligence trying to simulate how the brain really works and computes mathematical problems. Eastern Vedic Mathematics is a superior mental, one-line system of arithmetic that will enhance the current 'Western' mathematics in school curriculae. Learn why you no longer need to know your Multiplication Tables of 6, 7, 8 and 9. Its brilliant.

The Vedic Scholars from thousands of years ago invented the concept of zero. Without 'Zero', Jain asserts, we would not have computers nor rockets to the moon. It was the greatest discovery for the advancement of the technological era that we are in. Currently, mathematicians have no known way of determining say the square root of 38. Jain shows how it can be effortlessly done to 6 decimal places in one line of working out. Then, unbelievably, he mentally calculates the cube root of numbers in the hundreds of thousands! Stunning, but easily explained. This video is richly illustrated with relevant diagrams and photos of the Masters and makes learning easy by being able to stop and start the lecture and replay certain parts all in the comfort of your lounge-room. There will be a series of 'studio produced' 1 hour videos following this designed as a step-by-step

Teacher-Training Course for the serious student wishing to become a human bio-calculator.

– 8 –
THE BOOK OF PHI,
subtitled: The Living Mathematics of Nature Volume 1

ISBN: 0-9594180-3-2, RRP = $40, pub 2002, pp 172, gms 510, A4, sb, cov Dark Green, hw and computerised.

DESCRIPTION:

The Divine Proportion, aka the Sacred Cut, aka the Golden Mean or the Phi Ratio is highly illustrated in this fascinating book that provides insights into Galactic and Atomic Mathematics. It is revealed in ancient temples and pagodas, in classical paintings like the Mona Lisa, in the cross-sectional view of D.N.A. molecules and transverse cuts of pawpaws and apple slices showing the packing of seeds in Phi-ve (5) Pointed Star arrangements. Phi is in the nesting of the 5 and only universal shapes that fit inside a sphere and having equal angles, lengths and faces: the 5 Platonic Solids which are the Keys to Alchemy, and obey a universal mathematical ratio determined by the famous Fibonacci Sequence: 0, 1, 1, 2, 3, 5, 8, 13, 21, 34, 55, 89, 144 etc. These numbers are noticed in the counter-rotating fields or spirals of the sunflower, pine cone and many other biological forms. This is the study of the Universe as reflected in all parts of the Human Body.

This is the Mathematics of Infinity: Phi (Φ) expressed as a principle or function or ratio rather than a number: "Unity is in relationship to 1.61803398 (the decimal part goes on forever without any detectable repetitious pattern) and has fascinated many cultures for millennia and been dubbed, by great ϕlosophers as a Transcendental Number, like Pi (Π), again having no handle to grab or understand its decimal path. Finally for the first time in print, a distinct handle or Pattern has been found and explained and is revealed in Chapter 7 of this book: 'Phi Decoded. Secret Formula Revealed'. The clue is related to the external angle of the Pentagram being 108 degrees. This corresponds directly with the Mantra for Enlightenment, chanted 108 times, known as the Gayatri Mantra which has a sanskrit value of 108 and counted with the 108 rosary beads. Its is a sonic code. All Mathematics is meaningless without Sound Vibration. The Shape corresponding to the 108 phi Code is the Dodecahedron (having 12 pentagonal faces with angles of 108 degrees!) which either needs to be constructed within our Hearts and or fractally built etherically around us in our Meditations. This Phi Ratio is part of our Ascension Process and is an important clue to accessing our Multi-Dimensionality.

– 9 –
IN THE NEXT DIMENSION
THE BOOK OF PHI, Volume 2
The 17th Vedic Mathematics Sutra

ISBN: 0-9594180-2-4, pub. 2003 RRP $50, pp 190+, size A4, gms 700 approx, cov Light Green, sb, computerised text only.

DESCRIPTION:

Only Vedic Mathematics can decipher the 108 Phi Code: The Fibonacci Numbers, which generate Phi 1:1.618, has a distinct recursion of 24 digits which is intelligently composed of 12 Pairs of 9, summing to 108. This book validates Base 12 as the Galactic Maths and Base 10 as the Earth Maths necessary for quick mental calculations as in Vedic maths. The 108 Phi Code is embedded in the Dodecahedron which is the 3-Dimensional Form of the Pentagon. This book reveals the discovery of Jain's original formulae in the field of sacred geometry eg: the sequence of Magic Square Constants: 15, 34, 65, 111, 175, 260 etc exist in the next dimension as triplets of Tetrahedral Numbers. Jain has also discovered a more efficient way of expressing nature's number, the exponential "e" based on the powers of numbers relating to Vedic Sutras. The Powers of Numbers like 'x' squared and 'x' cubed etc really relate to different dimensions: ("e" = 2.71828182845904523 infinite, relates to populations of bacteria, and human populations etc). This book also explores shapes in 2D, 3D, 4D etc giving the reader a greater insight into the 5 Platonic Solids and the 14 Archimedean Solids as clues to Atomic Structure.

Jain also shows his original work with 2D Magic Square Patterns resembling identically the shadows of the atomic structure of 3D crystals like Berylium. Jain explains how Shape is Memory. Shapes, like crystals, know how to Store Memory. Think of the silicon chip whose atomic structure is the Star Tetrahedron or two inter-digitating triangular based pyramids. It knows how to store and memorise great and almost unimaginable quantities of data! It is also the 3D form of the 2Dimensional Star of David. This book is the result of 25 years of investigation into number patterns mysteriously tapping into nuclear geometries. Even if you don't read it and merely look at the diagrams on each page you will be inspired.

– 10 –
DECALS

DESCRIPTION:

These are adhesive window transparencies, silkscreened with 6 colour overlays and are available as Magic Square Patterns. When placed on the inside of a glass window, the sun projects the geometry into the room as coloured light. This can be projected onto the body for healing purposes. Different Magic Square Numbers are traditionally attributed to the seven known planets, and their specific Magic Sums or Magic Square Constants, like 15, 34, 65, 111 and 175 etc were secret numerical and ecclesiastical codes for the many multi-Cultural Names of God, YHWH, Allah, Messiah, Jesus, Adonai, Abraxas, etc.

You can choose from 2 sizes from the following 3:

.A. THE MAGIC SQUARE OF 4 X 4 (Jupiter) Small size (135mm diameter), has a magenta circular background. $8.

.B. THE MAGIC SQUARE OF 8 X 8 (Mercury) Small size (135mm diameter), has a blue circular background. $8.

.C. THE MAGIC SQUARE OF 4 X 4 (Jupiter) Large Size (310mm diameter), has a yellow circular background. Same pattern as the smaller one. $16.

– 11 –
THE GOSPEL
OF THE HOLY TWELVE
aka THE GOSPEL OF THE PERFECT LIFE

edited by: A Disciple of the Master.
reprinted 1974 by Health Research, California.
RRP $40. pp 242, size A4, gms 410, sb.

DESCRIPTION:

This rare Essene Bible was found in an urn in Tibet in a Buddhist monastery where it had been laying for 1800 years until it was recently rediscovered and translated from the ancient Aramaic (the language Jesus spoke) into English. Of the 4 translators, one of them was the famed Emmanuel Swedenborg (died 1772), the Swedish mystic and seer.

My particular interest in this mms is for 4 reasons:

1: I first discovered this book at the Adyar bookshop, Sydney, around the time of 1977 when I was 20 years of age and it was protruding from a top shelf. As I jumped to push the book back in place, I suddenly decided to grab it. I normally would not be standing in front of Christian material. This book changed my life.

2: As I flipped through the pages, this book had many images of Magic Squares of 3x3, 7x7 and 11x11 and sacred symbols like Pythagoras' Triangle, Egyptian Ankhs, Star of Davids, Alpha and Omega symbols, Ground Plan of the Christian Church, Labyrinth etc.

3: The Lord's Prayer, in Lection XIX, in the chapter: "Iesus Teaches Concerning Prayer" was the first book to honour both the Father and the Mother: It begins:

"OUR FATHER-MOTHER Who Art Above And Within: Hallowed Be Thy Sacred Name In Twofold Trinity . In Wisdom, Love and Equity Thy Kingdom Come To All...". I believe that the words "In Twofold Trinity" is a clear reference of the Phi Ratio which examines a Trinity of 3 parts into a specific twofold partitioning of 2 parts, one is the larger segment and the other is the smaller segment).

4: There is the most beautiful story about a man born blind from birth who Jesus heals. It is called: "The Examination of Him Who Was Born

Blind" (Lection LIV). This man becomes devoted to Iesus' teachings and asks many questions. It continues to describe the Divine Kingdom in terms of the Magic Square of 7x7 and is referred to as:
THE PARABLE OF THE SEVEN PALMS. From this material I have been inspired to create what I call THE THEATRE OF THE HOLY NUMBERS which is a creative enactment of this biblical scene.

The Magic Square of 11x11, like the 7x7 is important to the Essenes as the central crosses in both also add up to the Magic Sum or Constant of the Rows and Columns. In the Brief Commentary, the Magic Square of Eleven Perfected is subtitled: "Ecclesiae Militantis Sigilum", which means the magic Square of 11x11 is the Churches' Military Seal of Power or Defence.

This is a very rare book, with no ISBN; I have not seen another in print, nor met anyone that has heard of it. It appears to be from the same body of Essene works that documented the missing 18 years of Jesus' early life (as in The Aquarius Bible) deleted from the modern Bible, and is essential reading.

– 12 –
VEDIC MATHEMATICS
by: BHARATI KRSNA TIRTHAJI (1884–1960)
aka: The SHANKACHARAYA OF PURI.

ISBN: 81-208-0164-4, pub. by Motilal Banarsidass.
Soft Cover: RRP $40, pp 358, gms 610, Hard Cover: $50

DESCRIPTION:

This now famous book has launched a thousand websites. A highly intelligent and sophisticated system of Intuitive or Clairvoyant Mathematics based on 16 Sutras or Word Formulae. Shankacharaya was invited to America by Paramahansa Yogananda on a world tour in 1958 spreading the ancient message of Vedic Matheamatics. Originally Bharati Krsna (Bharati means India) had written 16 volumes or 16 books, a book per Sutra. Unfortunately these books were confirmed lost by a disciple, whilst Bharati Krsna was in America and he was asked by his disciples to quickly write an Introduction to Vedic Mathematics, as people in mainland India were now imitating this material and claiming psychic powers and not acknowledging the Vedic source. Bharati Krsna's memory was so astute he was able to dictate the whole contents of this book from mere memory. It is chiefly an Introduction to the lost 16 books and implies that there is so much more to come. Currently India has set up a World Vedic Mathematics Academy and are offering a large reward for the retrieval of these now priceless lost 16 books.

I (Jain) believe that my life's work constitutes a body or section of this Vedic material and I have summarised this into my latest book called IN THE NEXT DIMENSION (The Book of Phi, Volume 2, pub. July 2003) and is now known as The 17th Sutra, how a 2-Dimensional Magic Square is the shadow of a 3-Dimensional Crystal etc.

Shankacharaya's chief and current disciple is Maharishi Mahesh Yogi of Transcendental Meditation fame (who most people know as The Beatle's white-haired and bearded guru). Maharishi's spiritual school is churning out child geniuses like the 'Rain Man'. The Art of Mental Calculation is an important way of exercising the Brain's 'Mental Muscle'.

POSTAGE AND HANDLING CHARGES IN AUSTRALIA

Up to $40 order, add $6 for postage and handling
$40 to $120 order, add $10 for P and H
Over $120, free P and H. GST charges apply.

INTERNATIONAL ORDERS

1) International Bank drafts made in Australian Dollars only, or
2) Telegraphic Transfer.
No foreign currency cheques.
Air Freight or Sea Mail charges apply.

****

THERE ARE SPECIAL DISCOUNTS
FOR WHOLESALE ORDERS
AND FOR PURCHASING OF THE WHOLE SET OF 12 ITEMS.

IN MY NEXT BOOK
THE BOOK OF PHI, Volume 3

I will take you an another visual journey, deeper into the Phi Code and linking it directly to the DNA molecule.

We all know that the Binary Code or Doubling Sequence goes: $1 \rightarrow 2 \rightarrow 4 \rightarrow 8 \rightarrow 16 \rightarrow 32 \rightarrow$ etc but this is not the most efficient system, rather, when we specifically explore the **Vesica Piscis** and observe the count of the ever decreasing circles, we end up with a remarkable sequence that depicts perfectly the Phi Code's infinitely recurring 12 Pairs of 9. Here is the Sequence: $1 \rightarrow 2 \rightarrow 3 \rightarrow 6 \rightarrow 12 \rightarrow 24 \rightarrow$ etc.

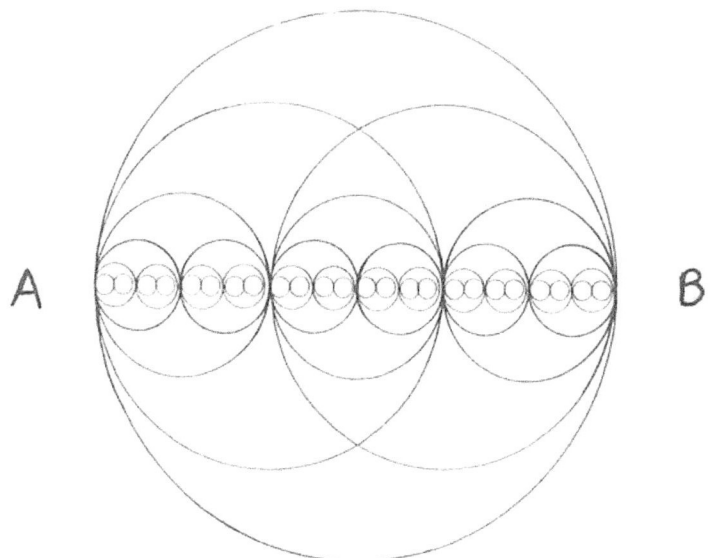

JAIN'S PHI-CODE VESICA PISCIS WITH PERIODICITY OF 24 AND WITH SEQUENCE OF $1 \rightarrow 2 \rightarrow 3 \rightarrow 6 \rightarrow 12 \rightarrow 24$

Notice that the original large Circle's Diameter AB is like a necklace of 24 glass beads in a linear sequence. If we call the center 'C', what happens if we slide the 12 circles of CB exactly over the 12 circles of AC. The overlapping of the two circles of the Vesica Piscis means that the Two has become One, again. This superimposition creates the 12 Pairs of 9:

① ① ② ③ ⑤ ⑧ ④ ③ ⑦ ① ⑧ ⑨
⑧ ⑧ ⑦ ⑥ ④ ① ⑤ ⑥ ② ⑧ ① ○

How can this linear sequence travel to **Infinity**? How does the last digit of the 24 sequence, the Zero, act as the vital link or bridge for the complementary pairs to self-replicate? If we don't count the zero as a number, is the Phi Code really 23 recurring numbers relating to the 23 contributing chromosomes in each human cell? Is it true that all biological proteins are pent molecules which are inherently Phi-based?

What 3-Dimensional solid embraces or embodies all of the sacred calendrical Time Code numbers like 12 months in a year and 12 years in Jupiter's cycle around the sun; 24 hours of the day; 30 days in a month; 60 years which allows 2 cycles of Saturn for 5 cycles of Jupiter, and 60 seconds in a minute? The answer is the **Dodecahedron** having **12** pentacle faces, a **24** periodicity in its Phi code, **30** edges and **60** plane angles.

<u>To be continued</u>: **THE PHI CODE + DNA.**

The IMPORTANCE of BASE 12, JAIN'S PETITION and PROPHESY.

Having read this book, literally or visually, you will come to appreciate that this startling discovery of the PHI CODE: the unravelling of a distinct periodicity of 12 Pairs of 9 Digits that recur and are embedded in the infinity of the Golden Measure 1:1.618033, attempts to place this new mathematical knowledge onto an accessible shelf which now must be labelled as **GALACTIC MATHEMATICS**.

Great teachers and pyramidologists were disheartened the day metric came in. Not only did we lose the original duodecimal Roman Fraction System based on twelfths, as in halves (1/2), thirds (1/3), quarters (1/4) and twelfths (1/12), we lost a money system based on 12 as in 12 pence to the Shilling, 12 items to the Dozen, and 12 oz (ounces) to the Troy.

Here is a footnote quote taken from Lawrence Blair's 'Rhythms of Vision', page 79, referring to John Michell, author of 'City of Revelation':

"John Michell raised an eloquent plea for the retention of Britain's now vanished duodecimal monetary system, on the grounds that it was the world's last link with the hierarchical modes of meaning behind money. With its fall to the merely decimal mode of energy, it appears that money acquires yet greater power in disproportion to the real rhythms which permeate society".

It is time to reclaim our natural identity with the 'Rhythms of Twelve' of life, now. If this is not achieved within one generation or 20 years of time, I see the children, their parents and concerned teachers out on the streets en-masse demanding an educational upheaval, the re-integration and reclamation of Base 12. When we apply 12 inches to the foot, we have 144 square inches to the squared foot. 144 is a harmonic of Light. Also, the measurement of the inch is based on a harmonic of the Earth's diameter, whereas the meter is arbitrary.

You must now ask yourself: "Why was I not educated about these important topics like the Phi Ratio, Platonic Solids, Vedic Mathematics and Magic Squares? I demand to know the truth of Mathematics, to tap into the memory stored in the Universal Language of its pictures, shapes, numbers and symbols. I understand now that all sacred sites, like the pyramids of Gizeh, Stonehenge etc are all based on divisions of 12, and without this Base 12 coding, I have no real tangible method of measuring these entities and therefore their measurements do not ring cranial bells meaning that there is a detectible sense of awe when we gaze at say the architectural beauty of the Parthenon, registering that its Phi-based measurements and my own personal measurements, like where the elbow bends, are also Phi-based. I understand that Base 12 will reveal the truth of these measurements and allow me to be in RESONANCE with these temples and astrophysical star maps that they are fractally diminutive of. I know that base 12 will allow the little micro picture of me to connect with the big macro picture of the stars. This theme of aesthetic self-similarity must be taught as an essential part of all Mathematical Enquiry.

I express no consternation towards Base 10, as shown in the body of knowledge known as Vedic mathematics, it is simply the intelligent Earth Mathematics that caters for lightning quick calculations. It is to be celebrated, alongside Base12.

I hereby sign the below dotted line casting my vote for the New Millennium Mathematics Curriculum Designers under the directorship of Jain of Australia and his Associates.
(Please sign below and send to the Author):

....
NAME

....
SIGNATURE   DATE

Jain Products Order Form

NEW! The Living Mathematics of Nature, Series – 5 DVD Set, pub: 2006

☐	Complete 5 DVD Set (Special! Save $25)	$200	240 gm
☐	Introduction to Ancient Knowledge (DVD #1 of 5)	$ 45	100 gm
☐	Vedic Mathematics (DVD #2 of 5)	$ 45	100 gm
☐	Magic Squares (DVD #3 of 5)	$ 45	100 gm
☐	The Divine Phi Proportion (DVD #4 of 5)	$ 45	100 gm
☐	3-Dimensional Geometry (DVD #5 0f 5)	$ 45	100 gm

Magic Squares

☐	The Book of Magic Squares, Volume 1 (M. Sqs. of 3, 4, 5, 6, 7)	$ 45	520 gm
☐	The Book of Magic Squares, Volume 2 (M. Sqs. of 8, 9, 10)	$ 45	520 gm
☐	The Book of Magic Squares, Volume 3 (M. Sqs. of 11, 12, 16)	$ 45	560 gm
☐	Join-The-Dots (adults and children coloring book. Pub:1999, 56 pp)	$ 30	160 gm
☐	Decal: The Magic Square of 4 X 4 (Jupiter) (Small, Magenta)	$ 10	20 gm
☐	Decal: Magic Square 8 X 8 (Mercury) (Small, Blue Background)	$ 10	20 gm
☐	Decal: Magic Square 4 X 4 (Jupiter) (Large, Yellow Background)	$ 20	100 gm
☐	The Gospel of the Holy Twelve (Essene Bible w. Magic Sqs. 210pp)	$ 45	410 gm

Vedic Mathematics

☐	The Magic of Nine in Vedic Mathematics. (Pub:1999, 124pp)	$ 45	390 gm
☐	The Vedic Mathematics Curriculum for the Global School. Part 1, Digital Sums. (Pub:2001, 142 pp)	$ 45	460 gm
☐	Jain Mathemagics Curriculum for the Global School. Part 2. Multiplication. (pub: 2005, 152pp)	$ 45	480 gm
☐	Vedic Mathematics for the New Millennium: DVD (Pub:2002)	$ 45	100 gm
☐	Dolphin Human Connections: Jain shows Vedic Maths (2006). 40 mins	$ 50	100 gm
☐	Vedic Mathematics by: Bharati Krsna Tirthaji **Soft Cov.** (355pp)	$ 45	460 gm
☐	Vedic Mathematics by: ibid. **Hard Cover.** (Pub:1965, 355pp)	$ 55	570 gm
☐	Calculations @ The Speed Of Light by Kranti Kiran (pub:2002, 104pp)	$ 45	230 gm

Sacred Geometry

☐	The Book of Phi, Vol 1 (The Living Maths of Nature. Pub:2002, 172pp)	$ 50	550 gm
☐	In The Next Dimension (The Book of Phi, Vol 2. Pub:2003, 188pp)	$ 50	580 gm
☐	Propositions on the Exact Value of Pi. Reddivari Sarva J. Reddy	$ 60	620 gm
☐	Peter's Projection Map. (Earth's true cartographic proportions)	$ 25	50 gm
☐	Dvd: Jonathan Quintin's "Visual Symphonies" Toroidal Animation	$ 45	100 gm

EarthHeart Jewellery

☐	Platonic Solids: Crystal Gift Set in wooden camphor boxes	$100	50 gm
☐	Heart Star: 24 carat gold plated stellated cuboctahedron	$150	50 gm
☐	Tantric Star: 24 carat gold plated Star tetrahedron	$130	50 gm
☐	Solar Star: 24 carat gold plated stellated dodecahedron	$150	50 gm

SUB-TOTAL: $ ....

Postage & Handling: $ ....

TOTAL: $ ....

Tax Invoice

Name: ....

Mailing address: ....

City/Town: .... State: .... Zip Code: ....

E-mail Address: ....

Home phone: .... Mobile: ....

Please choose your preferred method of payment:

☐ **Cash, Cheque or Money Order.**   Amount: ....
  Payable: JAIN F.R.E.E.D.O.M.S.

☐ **PayPal now Available!**   Amount: ....
  Email: jain@jainmathemagics.com

☐ **Direct Deposit**   Amount: ....
Acc Name: JAIN F.R.E.E.D.O.M.S. BSB: 032573   Acc Number: 171268

☐ **(circle) Visa / Mastercard**   Amount: ....

Card number ....

Name on Card: .... Exp. date: ....

AUSTRALIAN POSTAGE: Orders up to $50, add $8. Orders $50 to $120, add $12.

INTERNATIONAL ORDERS: $10 set up fee and add $10 per item. (Approx.) International Bank Draft in Australian Dollars. PayPal also available + Western Union. Cheques made out to JAIN F.R.E.E.D.O.M.S.

All Prices are listed in Australian Dollars. **All products available online!**

Date Of Order .... Dispatched ....

6	1	8
7	5	3
2	9	4

JAIN F.R.E.E.D.O.M.S. inc 9882763 trading as **JAIN MATHEMAGICS**
For **R**esearch **E**xpressing **E**ssential **D**ata **O**f **M**agic **S**quares
Phone: +61 02 6684 4409 Mobile: +61 0423 583 886 ABN 47 244 872 696
Email: jain@jainmathemagics.com Web: www.jainmathemagics.com
Address: 777 Left Bank Rd, Mullumbimby Creek, NSW, 2482, Australia

Distributed by Jain Mathemagics

The Living Mathematics Of Nature

5 DISC DVD SERIES

New DVD Series

Introduction to Jain Mathemagics
An introduction to the 4 topics below.
Explains each topic, providing good overview of work.

Vedic Mathematics
Rapid Mental Calculation.
No more calculators. Increases your memory power!

Magic Squares
Translating numbers into Atomic Art.
Teaches children pattern recognition.

The Divine Phi Proportion
Explores the Geometry of Flowers
identical to the Human Canon!

3-Dimensional Geometry
The 5 Platonic Solids
adored by Pythagoras and his community.

Distributed by JAIN MATHEMAGICS
777 Left Bank Road Mullumbimby
NSW 2482 Australia

Ph: (02) 6684 4409
Email: jain@jainmathemagics.com
www.jainmathemagics.com

Videos & Books Available Online...
www.jainmathemagics.com

Copyright © Jain Mathemagics 2005

Chapter 151

BACK COVER BLURB

JAIN MATHEMAGICS CURRICULUM FOR THE GLOBAL SCHOOL   part 2   MULTIPLICATION   2005

"THE FUTURE IN MATHEMATICS IS DOING IT IN YOUR HEAD" (by Jain: Mathematical Futurist).

- This is an Educational Workbook for Children and a Teacher's Resource Material for those interested in learning and teaching Rapid Mental Calculation. It increases the child's Memory Power and Confidence.
- This long-waited for book of Ancient Mathematical Short-Cuts excavates many hidden truths and vital properties in the playing field of Numbers.
- There is a rare and unique chapter called "HARMONIC STAIRWAY" from Jain's Dictionary of Numbers, for the first time in print. These choice Mathematical Plums are something that the student stumbles upon. They are working away solving calculations mentally, and when they check their answers in Chapter 2, they are instructed, in the section: DID YOU KNOW, to learn more about that particular number. But these are not any particular numbers, they are Anointed Numbers that our forebears held in high esteem.
- Beautifully and richly illustrated that the graphics alone educate. The student learns about other Families of Numbers, other numerical relatives and mathematical cousins that enrich their understanding of what mathematics is really about. It instills that sense of joy and wonderment that Pythagoras and Baudahayana knew.
- There is no error; it is an infallible bulletproof system, based on Unity Consciousness. The 16 "Threads" or Sutras that solve all known mathematical problems express The Law of Economy and The Path Of Least Resistance.
- As Jain Mathemagics becomes more globally acknowledged, it will help end the generational tyranny that has kept such knowledge in the dusty cupboard. This book is an organic pill that will prevent the slowly encroaching borgificiation of Mathematics.
- Since the turn of the bi-Millennium, there has been a global renaissance in the subject of "Sacred Geometry". You could summarize it as a fascination for the Language of Shape. This book invites seekers to truly incorporate SHAPE into their ability to perform mathematics, to apply this Language of Shape to the next octave of learning, not theoretically, but by practical use of using SHAPE to literally perform mental calculations within seconds.
- Often students come out of school unable to recover from deep mathematical wounds. They are mis-diagnosed as "un-intelligent" or "dumb", given drugs like Ritalin, told they are ADHD (which really means: Attuned Directly to Higher Dimensions). In fact, Dyslexic children are geniuses.
- **UNCOVER, RECOVER, DISCOVER.**

RAPID MENTAL CALCULATION
VEDIC MATHS for TEENS and ADULTS
MULTIPLICATION

$13 \times 14 = 13 + 4 / 3 \times 4 = 17 / 12 = 18 / 2 = 182$

$3^3 + 4^3 + 5^3 = 6^3$

Developing the Inner Mental Screen

$98 \times 97 = 98 - 3 / 2 \times 3 = 95 / 06 = 9,506$

BOOK 2

JAIN 108

from the series:

"JAIN MATHEMAGICS CURRICULUM FOR THE GLOBAL SCHOOL"

in the next dimension back cover blurb

- A distinct pattern has been found in the so called infinite, non-recurring decimal of Phi (Φ = 1:1.618033988), the living mathematics of Nature and Biology, the place where the elbow bends in relationship to the whole of the arm. Like Pi (Π = 3.1412, the relationship of the Circle's Circumference to its Diameter), mathematicians, for thousands of years, have been obsessed in finding internal patterning in these transcendental numbers. By stepping "In The Next Dimension" Jain has successfully cracked the Phi Code, by isolating a recurring pattern that has a distinct periodicity of 24 digits that re-align as 12 Pairs of 9, but it is not visible nor understandable in Western Maths, only in the Eastern Vedic Mathematics. The latter utilises a special Sutra or Formula that Digitally Compresses multi-digits to single digits, eg 144 = 1+4+4 = 9. His discovery and concept is officially known as the 17th Sutra.
- "In The Next Dimension" invites you to view a 2-D circle as a 3D sphere or as a 4D sphere which is a torus doughnut shape. Shape stores Memory, like crystals. The ultimate shape that embodies the Phi Code is the Dodecahedron having 12 pentacle faces, it being the 3D form of the 2D pentacle. Jain is petitioning for the reintroduction of the Base 12 number system as our original identification and resonance with the measurements of sacred sites like the Pyramids of Egypt and Stonehenge.
- This book explores 2 other major and original mathematical discoveries:

1– The Magic Square Constants (the Magic Sums like 15, 34, 65, 111 and 175) are encoded in the the Tetrahedral Number Series: (1, 4, 10, 20, 35, 56).

2– There is a better way to define "e" (=2.71828182845904523) another infinite non-recurring number like Pi and Phi that is the formula for the number of growth and decay in human or bacterial populations. Based on 'Similar Powers', it too glorifies the universal application of the Vedic Mathematics Sutras.

- Even if you don't read this book and merely look at the diagrams on almost every page, you will be inspired.

  Jain believes, through the highly visual content of this material, any child or adult will be triggered to remember the essential truths captured in the Universal Language of Pictures, Shapes, Numbers and Symbols.
- This is the Joy of the New Millennium Mathematics.

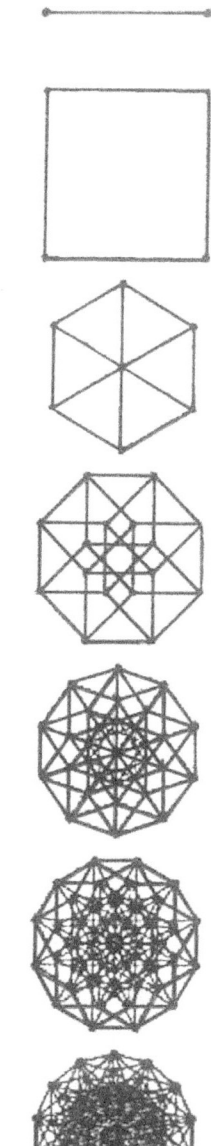

LEARN THE SECRETS OF NATURE

JainMathemagics.com

Jain's 5 Interactive Seminars will help you discover the beauty of numbers in Art, Science, Music and History.

THE ART OF NUMBER
Translating Numbers Into Art and Sacred Symbols

RAPID MENTAL CALCULATION
Vedic Mathematics Learn how to be a Human Bio-Calculator

MAGIC SQUARES
Translating Numbers Into Exquisite Atomic & Mandala Art

THREE-DIMENSIONAL GEOMETRY
of the 5 Platonic Solids Reveal Secrets of Atomic Structure

DIVINE PROPORTION
Living Mathematics of Nature and the Lost Secrets of the 108 Phi Code

Three Age Groups:
Juniors (9-12)
Teens
Adults

The 5 Day Seminars progress from:
~ Level 1 Introductory
~ Level 2 Advanced
~ Level 3 Teacher Trainer Courses